Catalytic Processes for CO₂ Utilization

Catalytic Processes for CO_2 Utilization

Editors

Markus Lehner
Juergen Karl
Reinhard Rauch

MDPI • Basel • Beijing • Wuhan • Barcelona • Belgrade • Manchester • Tokyo • Cluj • Tianjin

Editors

Markus Lehner
Chair of Process Technology
and Industrial Environmental
Protection, Montanuniversitaet
Leoben
Austria

Juergen Karl
Chair of Energy Process
Engineering, Friedrich-
Alexander-Universität
Erlangen-Nürnberg (FAU)
Germany

Reinhard Rauch
Engler-Bunte-Institute,
Karlsruhe Institute of
Technology
Germany

Editorial Office
MDPI
St. Alban-Anlage 66
4052 Basel, Switzerland

This is a reprint of articles from the Special Issue published online in the open access journal *Energies* (ISSN 1996-1073) (available at: https://www.mdpi.com/journal/energies/special_issues/catalytic_process_CO2_utilization).

For citation purposes, cite each article independently as indicated on the article page online and as indicated below:

LastName, A.A.; LastName, B.B.; LastName, C.C. Article Title. *Journal Name* **Year**, *Volume Number*, Page Range.

ISBN 978-3-0365-3603-3 (Hbk)
ISBN 978-3-0365-3604-0 (PDF)

© 2022 by the authors. Articles in this book are Open Access and distributed under the Creative Commons Attribution (CC BY) license, which allows users to download, copy and build upon published articles, as long as the author and publisher are properly credited, which ensures maximum dissemination and a wider impact of our publications.

The book as a whole is distributed by MDPI under the terms and conditions of the Creative Commons license CC BY-NC-ND.

Contents

Sascha Kleiber, Moritz Pallua, Matthäus Siebenhofer and Susanne Lux
Catalytic Hydrogenation of CO_2 to Methanol over Cu/MgO Catalysts in a
Semi-Continuous Reactor
Reprinted from: *Energies* **2021**, *14*, 4319, doi:10.3390/en14144319 . . . . **1**

**Michael Bampaou, Kyriakos Panopoulos, Panos Seferlis, Spyridon Voutetakis,
Ismael Matino, Alice Petrucciani, Antonella Zaccara, Valentina Colla, Stefano Dettori,
Teresa Annunziata Branca and Vincenzo Iannino**
Integration of Renewable Hydrogen Production in Steelworks Off-Gases for the Synthesis of
Methanol and Methane
Reprinted from: *Energies* **2021**, *14*, 2904, doi:10.3390/en14102904 . . . . **15**

**Philipp Wolf-Zoellner, Ana Roza Medved, Markus Lehner, Nina Kieberger and
Katharina Rechberger**
In Situ Catalytic Methanation of Real Steelworks Gases
Reprinted from: *Energies* **2021**, *14*, 8131, doi:10.3390/en14238131 . . . . **39**

Philipp Neuner, David Graf, Heiko Mild and Reinhard Rauch
Catalytic Hydroisomerisation of Fischer–Tropsch Waxes to Lubricating Oil and Investigation
of the Correlation between Its Physical Properties and the Chemical Composition of the
Corresponding Fuel Fractions
Reprinted from: *Energies* **2021**, *14*, 4202, doi:10.3390/en14144202 . . . . **61**

Katrin Salbrechter and Teresa Schubert
Combination of b-Fuels and e-Fuels—A Technological Feasibility Study
Reprinted from: *Energies* **2021**, *14*, 5250, doi:10.3390/en14175250 . . . . **75**

Article

Catalytic Hydrogenation of CO₂ to Methanol over Cu/MgO Catalysts in a Semi-Continuous Reactor

Sascha Kleiber, Moritz Pallua, Matthäus Siebenhofer and Susanne Lux *

Institute of Chemical Engineering and Environmental Technology, Graz University of Technology, NAWI Graz, Inffeldgasse 25C, 8010 Graz, Austria; kleiber@tugraz.at (S.K.); palluamoritz1@gmail.com (M.P.); m.siebenhofer@tugraz.at (M.S.)
* Correspondence: susanne.lux@tugraz.at

Abstract: Methanol synthesis from carbon dioxide (CO_2) may contribute to carbon capture and utilization, energy fluctuation control and the availability of CO_2-neutral fuels. However, methanol synthesis is challenging due to the stringent thermodynamics. Several catalysts mainly based on the carrier material Al_2O_3 have been investigated. Few results on MgO as carrier material have been published. The focus of this study is the carrier material MgO. The caustic properties of MgO depend on the caustification/sintering temperature. This paper presents the first results of the activity of a Cu/MgO catalyst for the low calcining temperature of 823 K. For the chosen calcining conditions, MgO is highly active with respect to its CO_2 adsorption capacity. The Cu/MgO catalyst showed good catalytic activity in CO_2 hydrogenation with a high selectivity for methanol. In repeated cycles of reactant consumption and product condensation followed by reactant re-dosing, an overall relative conversion of CO_2 of 76% and an overall selectivity for methanol of 59% was obtained. The maximum selectivity for methanol in a single cycle was 88%.

Keywords: CO_2 hydrogenation; methanol; caustic MgO; bifunctional catalyst

Citation: Kleiber, S.; Pallua, M.; Siebenhofer, M.; Lux, S. Catalytic Hydrogenation of CO₂ to Methanol over Cu/MgO Catalysts in a Semi-Continuous Reactor. *Energies* **2021**, *14*, 4319. https://doi.org/10.3390/en14144319

Academic Editor: Markus Lehner

Received: 10 June 2021
Accepted: 13 July 2021
Published: 17 July 2021

Publisher's Note: MDPI stays neutral with regard to jurisdictional claims in published maps and institutional affiliations.

Copyright: © 2021 by the authors. Licensee MDPI, Basel, Switzerland. This article is an open access article distributed under the terms and conditions of the Creative Commons Attribution (CC BY) license (https://creativecommons.org/licenses/by/4.0/).

1. Introduction

The steadily increasing carbon dioxide (CO_2) concentration in the atmosphere demands reduction of CO_2 emissions and necessitate CO_2 mitigation strategies [1]. A potential approach is the carbon capture and utilization (CCU) strategy, in which CO_2 is captured from large industrial contributors, such as the iron and steel industries and cement production, and converted into value-added chemicals. A promising product is the bulk chemical methanol (CH_3OH), which is used as solvent for paints, plastics, and adhesives, as feedstock for the production of numerous chemicals, such as formaldehyde, ethylene, propylene, methyl tertiary-butyl ether, and acetic acid, as fuel additive, and for fuel cell applications. Due to its higher performance, lower emissions and lower flammability compared to gasoline, methanol is classified as an alternative to conventional fossil-based fuels [2–10]. Methanol can be used as an energy carrier to store excess energy from wind and solar power plants at peak production times. Excess electric energy is converted into chemical 'hydrogen-fixed energy' by electrolysis of water, and consecutive synthesis of methanol via CO_2 hydrogenation improves the energy density of H_2-based energy carriers by one order of magnitude [11]. Methanol easily releases H_2 by steam reforming, it is, therefore, highly feasible for fuel cell powering [12]. Gas turbines have been shown to successfully run on methanol, which can be used to provide electricity in remote regions [13].

The state-of-the-art technology of methanol synthesis is based on the hydrogenation of syngas, a mixture of carbon monoxide (CO), CO_2, and H_2. The most common composition for syngas to methanol synthesis is given in Equation (1) [14].

$$\frac{n_{H_2} - n_{CO_2}}{n_{CO} - n_{CO_2}} = 2 \qquad (1)$$

Syngas is mainly produced by steam reforming of natural gas (CH_4) according to Equations (2) and (3). The hydrogenation reactions of CO (Equation (4)) and CO_2 (Equation (5)) are exothermic reactions. In both reactions, the total number of moles decreases. CO_2 is partially reduced to CO via the endothermic reverse water–gas shift reaction (RWGS, Equation (6)).

$$CH_4 + H_2O \rightleftharpoons CO + 3\,H_2 \qquad \Delta H_{R,\,298\,K} = 206\ \text{kJ mol}^{-1} \qquad (2)$$

$$CH_4 + 2\,H_2O \rightleftharpoons CO_2 + 4\,H_2 \qquad \Delta H_{R,\,298\,K} = 165\ \text{kJ mol}^{-1} \qquad (3)$$

$$CO + 2\,H_2 \rightleftharpoons CH_3OH \qquad \Delta H_{R,\,298\,K} = -91\ \text{kJ mol}^{-1} \qquad (4)$$

$$CO_2 + 3\,H_2 \rightleftharpoons CH_3OH + H_2O \qquad \Delta H_{R,\,298\,K} = -50\ \text{kJ mol}^{-1} \qquad (5)$$

$$CO_2 + H_2 \rightleftharpoons CO + H_2O \qquad \Delta H_{R,\,298\,K} = 41\ \text{kJ mol}^{-1} \qquad (6)$$

According to the principle of Le Chatelier, a low temperature and high pressure favor methanol synthesis. However, due to the chemical inertness and thermodynamic stability of CO_2 elevated reaction temperatures are necessary to activate CO_2 and facilitate methanol synthesis [15]. With industrial standard reaction conditions at temperatures of 523–573 K and a pressure of 5–10 MPa over $Cu/ZnO/Al_2O_3$ catalysts, a carbon conversion of 50–80% can be achieved [16]. The overall yield of methanol is limited by the thermodynamic equilibrium (Figure 1a).

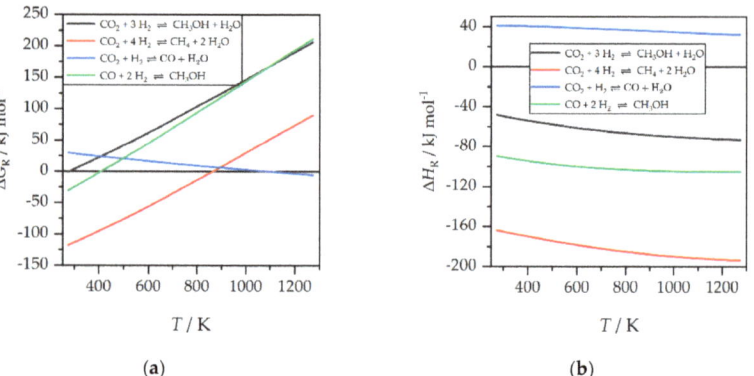

Figure 1. (**a**) Standard Gibbs free energies of reaction (ΔG_R^0) for carbon dioxide (CO_2) hydrogenation to methanol and methane, reverse water gas shift reaction (RWGS), and carbon monoxide (CO) hydrogenation to methanol; (**b**) Enthalpies of reaction (ΔH_R) for CO_2 hydrogenation to methanol and methane, RWGS, and CO hydrogenation to methanol; data calculated with HSC Chemistry 8 [17].

Figure 1a shows the basic problem of methanol synthesis by CO_2 hydrogenation. The standard Gibbs free energy of reaction (ΔG_R^0) is positive throughout the whole temperature range, suggesting specific operation conditions for successful synthesis by making use of the Le Chatelier principle, and by removing the reaction products during synthesis. According to Figure 1b, the economic success of methanol synthesis, of course, depends on sophisticated heat energy management, for example, by transferring the enthalpy of reaction (ΔH_R) from synthesis to distillative methanol/water separation.

In order to shift the carbon source for the synthesis of methanol from fossil-based fuels to CO_2 from industrial processes, it is crucial to provide cheap and robust catalysts for direct hydrogenation of CO_2 to methanol. Industrial catalysts for syngas conversion to methanol are not as effective in CO_2 hydrogenation [18].

In the scientific literature, there is still disagreement regarding the reaction mechanism of CO_2 hydrogenation to methanol. Some researchers postulate a one-step direct hydrogenation of CO_2; others report a two-step hydrogenation process via CO. Direct hydrogenation of CO_2 can be depicted from Equation (5) [9,19,20]. Based on C^{14} tests, it is reported that methanol is primarily produced from CO_2, while CO is oxidized to CO_2 according to the water–gas shift reaction (reverse Equation (6)) [3,21–23]. Increasing the CO_2 content in the syngas up to 30 mol% improves the energy balance and methanol yield [24]. Higher CO_2 concentrations seemingly inhibit the methanol synthesis, as CO_2 is converted to CO by the RWGS reaction. The by-product water shifts the equilibrium of Equation (5) towards the reactants and deactivates the catalyst by inhibiting the active sites [3,15,25]. Other researchers discuss a two-step hydrogenation mechanism, in which CO_2 is reduced to CO first according to the water–gas shift reaction, and CO is then converted to methanol according to Equation (4) [26,27].

Various catalysts for methanol synthesis from CO_2 have been developed and intensively investigated over the last decades. The main influencing factors for the catalytic activity, stability, and selectivity of the catalysts are the process conditions, the preparation method and the choice of the catalytically active material, the catalyst carrier material, and the use of promotors. The target of optimum process conditions, such as temperature, pressure, feed gas composition and flow rate, the amount of catalyst, and continuous or batch operation mode is controlled by the thermodynamics of the reaction. The choice of carrier material, additional promotors, and the preparation method affects catalyst parameters such as particle size, surface area, metal distribution, acidity and basicity, temperature and pressure stability. In general, catalysts for methanol synthesis by CO_2 hydrogenation can be categorized as follows: Cu-based catalysts, noble metal-based catalysts (Pd, Pt), oxygen-deficient catalysts (In_2O_3), and bimetallic catalysts (Ni-Ga, Au-Ag) [7,8].

Cu-based catalysts have attracted research interest and they are already industrially applied, mainly with the carrier Al_2O_3 and the promotor ZnO. In Cu-catalyzed synthesis of methanol by CO_2 hydrogenation, the nature of the carrier material has a pronounced effect on the reaction [28]. The catalytic activity linearly correlates with the metallic Cu^0 surface [29,30], indicating that the reaction takes place at the metallic Cu^0 surface [31]. Several studies have shown that the admixture of MgO as an additional promotor increases CuO dispersion, the metallic Cu^0 BET surface area, and the active basic sites for improved CO_2 and H_2 adsorption [31–40].

While Cu-based catalysts on MgO carrier without additional promotors have rarely been described for methanol synthesis from CO_2 so far, the bifunctional catalytic effect of catalysts with MgO carrier material is well described in CO_2 hydrogenation to methane. In CO_2 methanation with Pd/MgO/SiO_2 catalysts it was found that MgO initiates the reaction through adsorbing CO_2 molecules and thus forming magnesium carbonate on the surface. The reaction proceeds with atomic hydrogen provided by Pd. Atomic hydrogen is essential in hydrogenation of magnesium carbonate to methane. After desorption of methane the carbonate regenerates through gaseous CO_2. The Pd/MgO/SiO_2 catalyst was calcined at 823 K [39]. Loder et al. [40] investigated the reaction kinetics of CO_2 methanation with bifunctional Ni/MgO catalysts. They developed a kinetic model based on the Langmuir–Hinshelwood reaction mechanism considering H_2 adsorption and dissociation and CO_2 adsorption on the catalyst to take the bifunctional catalytic action of the catalyst into account.

MgO (also called magnesia) may be grouped in three grades depending on the calcination temperature: (i) caustic MgO, (ii) sintered MgO, and (iii) fused MgO. Caustic MgO is formed when $Mg(OH)_2$ or $MgCO_3$ is heat treated slightly above the decomposition temperature. It has a very high caustic reactivity in terms of neutralization rate with HCl. Depending on the calcination temperature, light-burnt (1143–1273 K) and hard-burnt (1823–1923 K) MgO may be distinguished. The caustic reactivity of MgO decreases with increasing calcination/sintering temperature. Sintered MgO (also called dead-burnt MgO) is calcined at temperatures of 1673–2273 K. It shows a high heat storage capacity and a high thermal conductivity but low caustic reactivity. Fused MgO is crystalline magnesium oxide, formed above the fusion point of MgO (3073 K). Its strength, abrasion resistance, and chemical stability are superior compared to sintered MgO. In reducing atmosphere, it is stable up to 1973 K. The chemical properties of MgO strongly depend on the calcination temperature and duration. In general, with increasing calcination temperature and/or duration the specific surface area and the distortion of the crystal lattice decrease, and the particle size increases, resulting in decreasing reactivity of MgO [41,42].

However, to the best of our knowledge, the effect of calcination temperature and duration on the catalytic effect of catalysts with MgO as carrier material or promotor has not been investigated so far. From previous studies [40] it has become evident that the caustic behavior of MgO and its adsorption capacity for CO_2 plays a fundamental role in CO_2 hydrogenation with catalysts based on MgO as carrier material. Yang et al. [43] studied the CO_2 adsorption capacity of MgO-based adsorbents calcined at different temperatures. It was shown that with increasing calcination temperature up to 823 K the adsorption performance got better, while above 873 K, it started to decrease. While still being in the range of light-burnt caustic MgO, higher calcination temperature led to a reduction of the BET surface area and eliminated part of the intergranular porous structure, hindering diffusion of CO_2 in the particles and decreasing the adsorption capacity.

In CO_2 hydrogenation to methanol water is formed as by-product. Many catalysts suffer from deactivation by water. Salamão and Pandolfelli [44] investigated the hydration-dehydration behavior of MgO sinter. They used partially hydrated sintered MgO and studied the effect of the calcination temperature (383–1173 K) on its reactivity. Partially hydrated MgO sinter is characterized by a thin film of $Mg(OH)_2$ on the surface. When calcining at moderate temperatures of 623–873 K, the $Mg(OH)_2$ layer totally decomposes but the original structure of MgO is not regained. For calcination above 873 K, the initial structure of MgO is recovered, but surface area and reactivity will deplete. These findings clearly show the pronounced impact of the calcination conditions on MgO-based catalysts.

The gap in detailed consideration of the effect of MgO preparation on the catalytic activity initiated the investigation of Cu/MgO catalysts in this study. The bifunctional catalyst Cu/MgO suffices the requirements of simple preparation, activity at moderate reaction conditions, and low technological demand for recycling in blast-oxygen furnaces in the copper industry.

This paper provides first results with Cu/MgO catalysts in methanol synthesis from CO_2 in a semi-continuous tank reactor. MgO was prepared from $MgCO_3$ at low temperature to provide high caustic reactivity.

2. Materials and Methods

2.1. Materials

For preparation of the Cu/MgO catalyst, copper(II) nitrate trihydrate ($Cu(NO_3)_2 \cdot 3\,H_2O$, ≥99.5%, p.a. ACS), granulated spherical MagGran© ($4\,MgCO_3 \cdot Mg(OH)_2 \cdot 4\,H_2O$, Ph. Eur., Magnesia AG, Switzerland) with a particle size distribution of 0–8 wt% < 150 µm, 0–15 wt%: 150–250 µm, 55–80 wt%: 250–600 µm, and deionized water were used. H_2 (99.999%), CO_2 (99.998%), and nitrogen (N_2, 99.999%) supplied by AirLiquide were used for the hydrogenation experiments.

The Cu/MgO catalyst with a mass fraction of 38 wt% Cu with respect to the mass of the MgO carrier material was prepared via wet impregnation. The method was adapted from Loder et al. [40] from the preparation of bifunctional Ni/MgO catalysts for CO_2 methanation. Catalyst preparation consisted of four steps:

1. Calcination: To prepare the catalyst carrier MgO, MagGran© granulate was calcined in air in a muffle furnace (Heraeus M110) for five hours under mild conditions at 723 K followed by two hours at 823 K (Equation (7)).

$$4\,MgCO_3 \cdot Mg(OH)_2 \cdot 4\,H_2O \rightleftharpoons MgO + 4\,CO_2 + 5\,H_2O \qquad (7)$$

2. Impregnation: The calcined MgO granulate (13 g, white) was impregnated with 0.25 dm³ of an aqueous copper(II) nitrate solution (c_{Cu} = 35 g dm^{-3}) in a water cooled

flask under constant stirring. After two hours, the impregnated catalyst precursor (blue) was filtered off and dried overnight in a drying furnace at 303 K.

3. Thermal decomposition: The impregnated dry catalyst precursor was calcined in the muffle furnace for one hour at 423 K followed by five hours at 723 K (Equation (8)). Calcination resulted in a change of color from blue to black.

$$Cu(NO_3)_2 \cdot 3\,H_2O \rightleftharpoons CuO + 2\,NO_2 + 3\,H_2O \quad (8)$$

4. Reduction (catalyst activation): To generate the catalytically active Cu^0 sites, the calcined CuO/MgO precursor was treated in H_2 atmosphere for 3.5 h in the tank reactor that was also used for the hydrogenation experiments (Equation (9)). The activation of the catalyst was performed at the reaction conditions of the CO_2 hydrogenation experiments at 573 K and 5 MPa.

$$CuO + H_2 \rightleftharpoons Cu + H_2O \quad (9)$$

2.2. Experimental Setup and Procedure

The experimental setup is depicted in Figure 2. It consisted of a semi-continuous tank reactor (BüchiGlasUster "Limbo350") equipped with an external recycle for the gaseous reactant stream and external condensation of condensable products. The volume of the reactor was 0.450 dm³. Gas recycle was performed by a Ziclón 04 gas circulation pump from Fink. The condensable products methanol and water were condensed in a heat exchanger (HE) at 275 K and collected in a condensate tank (0.1 dm³). The heterogeneous CuO/MgO catalyst precursor was placed in the reactor. The temperature of the reactor (HT1) and the riser to the heat exchanger (HT2) were controlled by an electrical heating system. The reactor was equipped with a wall cooling system. The temperature controller was operated by a process control unit based on LabView. The temperatures of the gas stream were measured by thermo-sensors inside the reactor (T1), before the heat exchanger (T2), inside the condensate tank (T3), and after the gas circulation pump (T4). The pressure was measured inside the reactor (P1) and the condensate tank (P2). The feed gas flow rates were adjusted by mass flow controllers (MFC). The recycle stream flow rate was measured by a mass flow meter (MFM). All temperatures, pressures, and mass flow rates were monitored and recorded. A needle valve was installed between the condensate tank and pump to withdraw gas samples during the experimental run. The samples were analyzed by micro gas chromatography (GC).

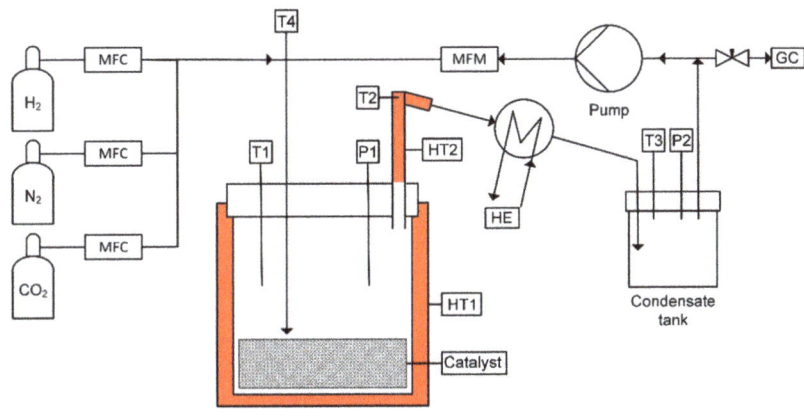

Figure 2. Experimental setup of the semi-continuous tank reactor with external recycle for the gaseous reactant stream and external condensation of condensable products in bench scale.

8 g of the CuO/MgO catalyst precursor with a particle size distribution of 200–600 µm was placed inside the reactor and activated with H_2 to reduce CuO to Cu according to Equation (9). After catalyst activation the reactor was depressurized to atmospheric pressure and kept under H_2 atmosphere. To start the hydrogenation experiment, the H_2:CO_2 feed gas ratio was adjusted to 3:1 (v:v) and the reactor was pressurized to 3 MPa at a constant feed gas flow rate of 600 scm^3 min^{-1}. A reference gas sample was taken before the reactor was heated to 573 K. At reaction temperature, the reactor was pressurized to 5 MPa. Then, the gas circulation pump was switched on providing a constant gas flow rate of 10 dm^3 h^{-1} over the whole experimental run. When the respective reaction conditions were reached, the experiment was operated for 48 h. Gas samples were withdrawn by opening the needle valve between the condensate tank and the circulation pump and filled into 20 cm^3 vials in three hour-intervals, starting 1.5 h after the experiment had been started. To determine the initial gas-phase composition, a reference sample was taken during the pressurization and heating phase. After having taken gas samples, the reactor was pressurized again to 5 MPa with the initial feed gas ratio of H_2:CO_2 of 3:1. The time span from pressurization until sampling was specified as an interval (cycle). The last sample was taken after a reaction time of 48 h, followed by depressurization and cooling of the reactor. Before the condensate tank was opened the reactor was purged with N_2 with a constant flow rate of 400 scm^3 min^{-1} for 30 min.

2.3. Analysis

Gas samples were analyzed by micro gas chromatography with an Agilent/Inficon microGC 3000 gas chromatograph equipped with two modules (module A and B). Each module consisted of a built-in column and a thermal conductivity detector (TCD). The injection temperature was 363 K in both modules. Module A had a 5 Å molsieve column with an inner diameter of 320 µm, a length of 10 m, and a thickness of 12 µm of the stationary phase. For pre-separation of the gases in backflush mode, a PLOT-U column was installed prior to the molsieve column. The PLOT-U column had an inner diameter of 320 µm, a length of 3 m, and a thickness of 30 µm of the stationary phase. This module was operated in backflush mode with the carrier gas argon. A column temperature of 373 K and pressure of 0.2068 MPa were used. The run-time was 120 s with additional 8 s of backflush time. With this module, H_2, N_2, CO, and CH_4 were detected. Module B had a PLOT-U column with an inner diameter of 320 µm, a length of 8 m, and 30 µm thickness of the stationary phase. It was operated with the carrier gas helium. A column temperature of 333 K and a pressure of 0.1724 MPa were used. The run-time was 120 s. With this module, CO_2 (and C_2H_6, C_2H_4, and C_2H_2) was detected.

The methanol concentration in the liquid product was determined by gas chromatography in accordance to the method described in [45]. The Shimadzu GC2010plus was equipped with a flame ionization detector (FID). A Zebron ZB WAXplus column with an inner diameter of 320 µm, a length of 60 m, and a thickness of 0.5 µm was used.

Calculations were based on the ideal gas law to quantify pressure changes by species formation and consumption, and to convert volumes at operation conditions into standard conditions (STP). The total reaction volume was obtained from the volume of the reactor, the volume of the condensate tank, and the volume of the piping. From the total reaction volume at standard conditions (V_{STP}), the volume fraction in the gas phase (φ_i) and the molar volume (v_i) the molar amount of each component (n_i) was calculated (Equation (10)). The relative conversion of CO_2 and H_2 (X_i) were calculated from the total amount of reactant at the beginning ($n_{i,0}$) and at the end ($n_{i,t}$) of an interval, and of the whole experimental run, respectively (Equation (11)). The yield of CO (Y_i) was calculated from the total molar amount produced ($n_{CO,t} - n_{CO,0}$) per mole of CO_2 fed to the reactor ($n_{CO2,0}$) in an interval and for the whole experimental run, respectively (Equation (12)). The total molar amount of methanol produced for each interval ($n_{CH3OH,t}$) was calculated with a carbon-based mass balance (Equation (13)). The selectivities for methanol and CO (S_i) were calculated from the relative conversion of CO_2 and the yield, respectively (Equation (14)).

$$n_i = \frac{V_{STP} \cdot \varphi_i}{v_i} \qquad (10)$$

$$X_i = \frac{n_{i,0} - n_{i,t}}{n_{i,0}} \qquad (11)$$

$$Y_i = \frac{n_{i,t} - n_{i,0}}{n_{CO_2,0}} \qquad (12)$$

$$n_{CH_3OH,t} = n_{CO_2,0} + n_{CO,0} + n_{CH_4,0} - n_{CO_2,t} - n_{CO,t} - n_{CH_4,t} \qquad (13)$$

$$S_i = \frac{n_{i,t} - n_{i,0}}{n_{CO_2,0} - n_{CO_2,t}} = \frac{Y_i}{X_{CO_2}} \qquad (14)$$

3. Results and Discussion

Figure 3 depicts the reaction temperature and pressure during the experimental run. The experimental run can be split into three phases: (i) heating and pressurization of the reactor to 573 K and 5 MPa, (ii) CO_2 hydrogenation over a period of 48 h, and (iii) cooling and depressurization of the reactor. In phase i, the reactor was filled with the reaction mixture in a stoichiometric ratio of $H_2:CO_2$ of 3:1 until pressure obtained a value of 3 MPa. After taking the reference sample, the reactor was heated to 573 K and finally pressurized with feed gas to 5 MPa. Phase ii started when the reaction conditions of 573 K and 5 MPa were reached. At that point the circulation pump was switched on and CO_2 hydrogenation was performed for 48 h. The temperature was kept constant over the whole experimental run. This reaction phase was characterized by repeated reaction intervals (15 in total). In each interval, the decreasing pressure over time at constant temperature is visible. This is characteristic for a volumetric contractive reaction and the condensation of the condensable products methanol and water. After each interval, a gas sample was taken and analyzed. The reactor was pressurized again to 5 MPa and the next interval started.

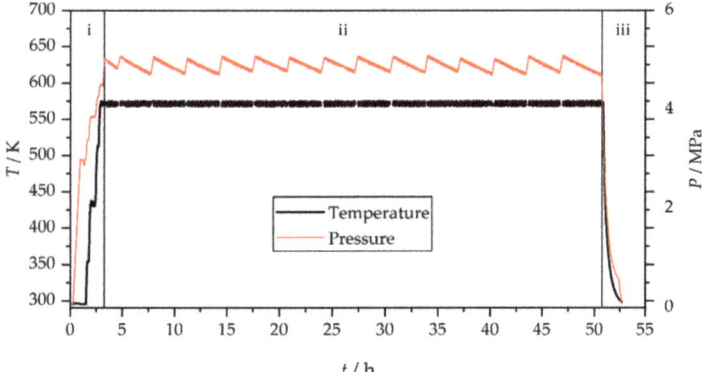

Figure 3. Reaction temperature and pressure during one experimental run, consisting of initial heating and pressurization of the reactor (phase i from 0 to 3 h), 48 h reaction phase characterized by repeated decrease of pressure due to reactant consumption and product condensation and re-pressurization (phase ii from 3 h to 51 h) followed by cooling and depressurization in phase iii.

According to the caustification conditions (five hours at 723 K followed by two hours at 823 K) the catalyst proved to be active with good selectivity for methanol. This may be dedicated to the caustic nature of MgO that shows high reactivity with respect to CO_2

adsorption. The conversion of the reactants H_2 and CO_2 as well as the formation of the by-product CO from the RWGS reaction can be monitored for each interval and over the total duration of the experiment. In Figure 4 the experimental results of the volume fractions of H_2, CO, and CO_2 in the gaseous reaction mixture are depicted over the reaction time. In addition to H_2, CO, and CO_2 negligible amounts of CH_4 were detected (below 1.5 vol%). C_2H_6, C_2H_4, or C_2H_2 has not been detected. When starting the experiment only H_2 and CO_2 were present in the gaseous reaction mixture. Within the first 24 h of the experiment, the amount of CO continuously increased, while the volume fractions of H_2 and CO_2 decreased. After this start-up phase, the gas-phase composition only denoted slight changes.

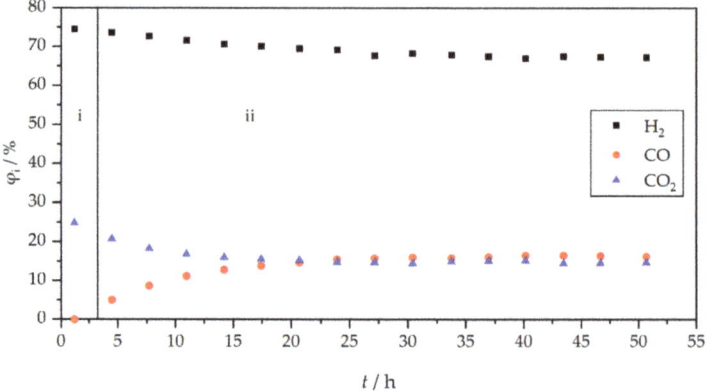

Figure 4. Volume fraction (φ_i) of the gaseous reaction mixture for CO_2 hydrogenation to methanol over the Cu/MgO (38 wt%) catalyst in the semi-continuous tank reactor with external recycle and condensation; operation conditions: 573 K and 5 MPa.

Figure 5 shows a cutout of the intervals 9, 10, and 11, exemplarily given for the volume fraction of CO in the gaseous reaction mixture. The points 1, 3, and 5 depict the volume fractions of CO at the end of the intervals 9, 10, and 11, respectively. These data represent measured volume fractions. As mentioned, after each interval and sample taking procedure, the reactor was pressurized again to 5 MPa with the feed gas H_2 and CO_2 in a stoichiometric ratio of 3:1. The volume fraction of constituents at the beginning of the following interval was different to the volume fraction of constituents measured at the end of the previous interval. The starting volume fractions at the beginning of the intervals 10 and 11 are represented by the points 2 and 4, respectively. The estimated trend in the concentration within one interval is depicted by the dashed line. The dot-dash' line forming the connection of points 1, 3, and 5 represents the measured gas-phase compositions.

From the gas-phase composition, the relative conversions of H_2 and CO_2 (Figure 6) and the yield (Figure 7) and selectivity (Figure 8) for methanol and CO can be obtained for each interval. The relative conversion of H_2 in the first interval, which lasted 1.5 h, was 8%. In the intervals 2 to 14, the relative conversion of H_2 fluctuated between 15% and 21%. As the last interval lasted for four hours to complete the 48 h of the experimental run, the relative conversion of the last interval was the highest with 23%. CO_2 conversion confirms the trend of the relative conversion of H_2. The lowest relative conversion of CO_2 of 22% was determined for the first interval, while the highest relative conversion of 32% was found in the last interval. In the intervals 2 to 14 the relative conversion of CO_2 was in the range of 23% to 29%.

The yield of the by-product CO showed a maximum value of 19% in the first interval. It steadily decreased to 4% at the end of the experiment. The yield of methanol had an opposite trend over the reaction time. The lowest value of 3% was obtained in interval

1. The maximum yield of methanol was 28%, recorded in the last interval. Referring to the positive trend of methanol yield it is concluded that the catalyst has still not obtained steady-state activity after 48 h of operation.

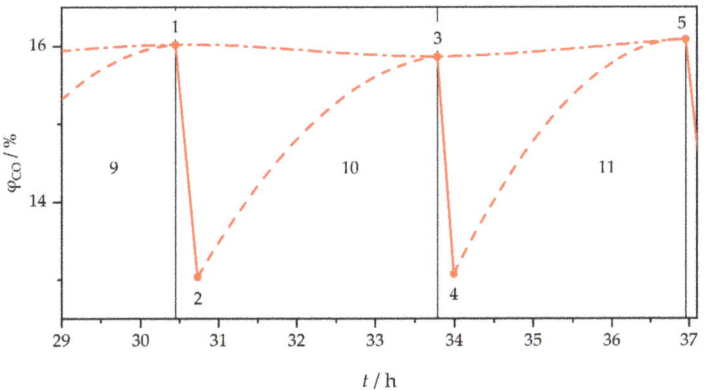

Figure 5. Cutout of the trend of the gas-phase volume fraction of CO for the intervals 9, 10, and 11. The points 1, 3, and 5 show the measured volume fraction at the end of the intervals 9, 10, and 11, respectively. The points 2 and 4 represent the calculated volume fraction at the beginning of interval 10 and 11, respectively. The estimated trend of the volume fraction within one interval is depicted by the dashed line. The 'dot-dash' line forming the connection of points 1, 3, and 5 represents the measured gas-phase volume fraction of CO.

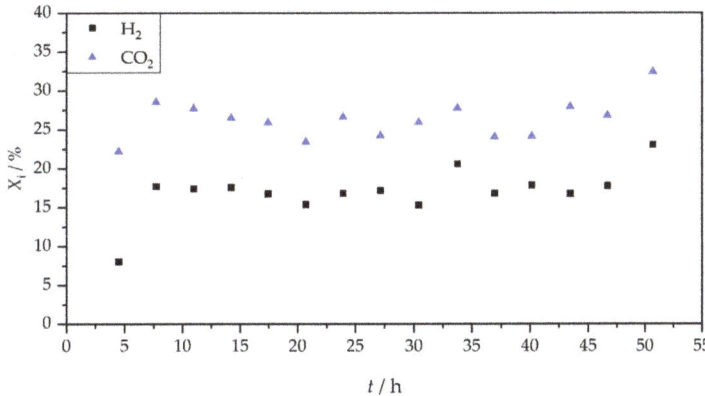

Figure 6. Relative conversion of H_2 and CO_2 for the different intervals during the experimental run.

The interval selectivities for methanol and CO resemble the yield of both products. In the first two intervals, CO is the preferred hydrogenation product. Within interval 3, this scenario changes towards methanol. The trend continues up to a maximum methanol selectivity of 88% in the last interval.

The bench-scale semi-continuous tank reactor was adapted according to state-of-the-art reactor configurations for industrial methanol synthesis [10]. As opposed to syngas-based industrial methanol synthesis, pure CO_2 was used as carbon source in this study. In the repeated cycles of reactant consumption and product condensation followed by reactant re-dosing, an overall relative conversion of CO_2 of 76% and a methanol selectivity of 59% were obtained. The calculated results based on the pressure loss during the experiment are consistent with the measured concentration of the liquid product at the end of the experiment with a deviation of ± 2%. This confirms the accuracy of the experimental procedure. Due to continuous product condensation both results are significantly higher than imposed by the thermodynamic equilibrium ($X_{CO_2,Eq}$ = 25% and $S_{CH_3OH,Eq}$ = 25% at 573 K and 5 MPa; calculated with HSC Chemistry 8 [17]).

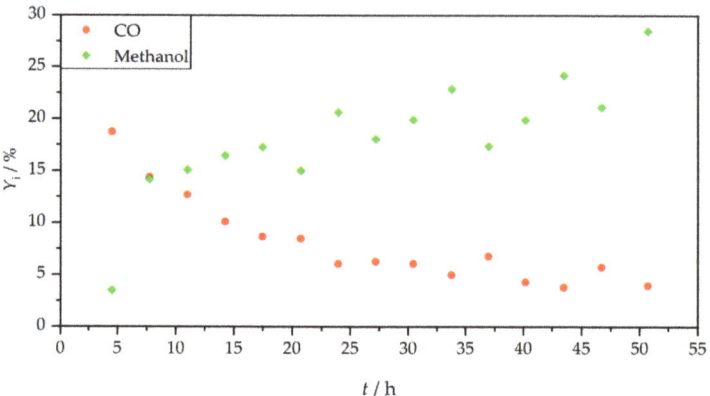

Figure 7. Interval yields of CO and methanol during the experimental run.

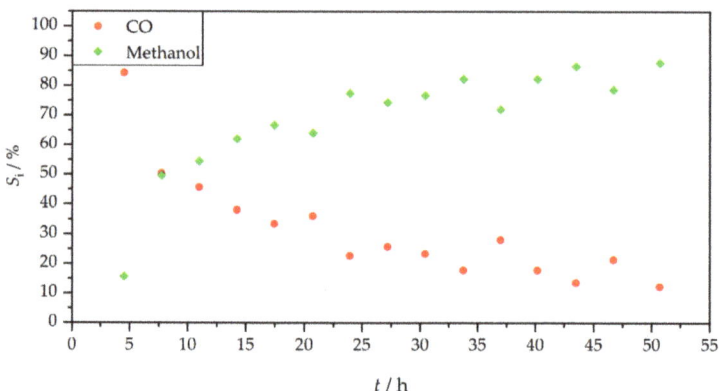

Figure 8. Interval selectivities of CO and methanol during the experimental run.

Ren et al. [32] investigated the promoting effect of ZnO, ZrO_2, and MgO on the activity of Cu/γ-Al_2O_3 catalyst. The admixture of the metal oxides increased the CuO dispersion, the Cu^0 surface area, and decreased the Cu^0 particle size. While the promoting effect of ZnO and ZrO_2 on CO_2 hydrogenation to methanol, when admixed separately, was marginal, simultaneous admixture of both oxides increased CO_2 conversion and methanol selectivity significantly. Further improvement was achieved with a Cu/ZnO/ZrO_2/MgO/γ-Al_2O_3 catalyst. The optimal temperature for catalyst activation was found to be the process temperature for CO_2 hydrogenation. Though lower activation temperatures resulted in the formation of smaller Cu^0 particles and the generation of a higher Cu^0 surface area, the catalyst particles seemed to agglomerate when the process temperature exceeded the activation temperature afterwards [32].

Dasireddy et al. [33,34,46] evaluated the effect of alkaline earth metal oxides (MgO, CaO, SrO, and BaO) on a Cu/Al_2O_3 catalyst for methanol synthesis from CO_2 and compared the results with commercially available Cu/ZnO/Al_2O_3 catalysts. The admixture of alkaline earth metal oxides enhanced the interaction between Al_2O_3 and CuO, which resulted in a weaker reducibility of CuO. The $Cu^+:Cu^0$ ratio and the Cu^0 surface area were higher for all alkaline earth metal oxide-containing catalysts compared to the $Cu/ZnO/Al_2O_3$ catalyst, and increased in the order of Ba < Ca < Zn < Sr < Mg. High $Cu^+:Cu^0$ ratio and high Cu^0 surface area were stated as decisive factors for high CO_2 conversion. Best results were obtained with $Cu/MgO/Al_2O_3$ catalysts, which showed an increased number of active sites for CO_2 and H_2 adsorption. Preparation conditions at pH = 8 further increased the positive effect of MgO-promoted catalysts. The catalytic performance even exceeded the commercially available benchmark catalyst HFRI20 and LURGI catalysts [33,34,46].

The oxidation state of Cu-species in perovskite-type catalysts (La-Cu-Zn-O) prepared with various promotors was the focus of the study of Zhan et al. [35]. A separate admixture of Ce_2O_3, MgO, and ZrO_2 promoters on perovskite-type catalysts improved the selectivity for methanol compared to the unpromoted catalyst. The highest methanol selectivity was obtained with MgO-promoted catalyst. The higher selectivity was assigned to an increased concentration of caustic sites, higher Cu dispersion, and a special $Cu^{\alpha+}$ species, which was different to Cu^0, Cu^+, and Cu^{2+} [35].

Liu et al. [36] investigated the influence of MgO-promoted Cu/TiO_2 catalysts. Admixture of MgO increased the number and strength of caustic sites, but decreased the reducibility of CuO, which was found beneficial for methanol selectivity [36].

Zander et al. [37] compared a Cu/MgO catalyst derived from Cu and Mg (molar ratio of 80:20) nitrate solutions via co-precipitation to a classical malachite-derived Cu/ZnO catalyst. They investigated hydrogenation with various feed gas compositions; pure CO_2, CO_2/CO mixture and pure CO feed gas stream, at 503 K and 3 MPa. Calcination of the catalyst precursor was carried out in air at 603 K. When pure CO_2 and mixed CO_2/CO feed gas streams were hydrogenated, the Cu/ZnO catalyst showed a much higher activity than the Cu/MgO catalyst. The Cu/MgO catalyst remained almost inactive in methanol synthesis and catalyzed the reverse water–gas shift reaction instead. The results indicated that the rate of methanol synthesis was not only a function of the exposed Cu surface area. The low activity of Cu/MgO in CO_2 hydrogenation was explained by the absence of a strong metal–carrier interaction in the investigated temperature regime.

Nielsen et al. [31] also studied the catalytic effect of Cu/MgO (20 wt%) catalysts in hydrogenation of pure CO_2 and pure CO feed gas streams. The catalysts were prepared via precipitation. No information was given about the calcination conditions. Under the applied hydrogenation conditions (523 K and 5 Mpa), Cu/MgO showed high catalytic activity in CO hydrogenation and only little activity in CO_2 hydrogenation. In CO_2 hydrogenation the relative CO formation rate was five times as high as the relative methanol formation rate. It was concluded that on Cu/MgO, the CO-pathway is much faster, arising from a bifunctional mechanism. They concluded that the facile CO hydrogenation on Cu/MgO proceeds via formate intermediates at the metal/oxide interface. The formate intermediates arise from CO that is inserted into a caustic OH-group from the oxide. This step is followed by Cu-assisted hydrogenation of formate to methanol. In the presence of CO_2 carbonates are formed and replace the formate species and thus show an inhibiting effect on methanol synthesis from CO. The catalytic effect of Cu/MgO catalysts on low temperature methanol synthesis from syngas with ethanol as promotor was found to be beneficial by Yang et al. [38].

It is assumed that the mentioned Cu-based catalysts [31–40] with MgO carrier/promotor were prepared at moderate temperatures providing reactive caustic MgO. It has not been explicitly mentioned in the corresponding papers. Cu/γ-Al_2O_3 catalysts modified by ZnO, ZrO_2 and MgO were prepared via impregnation method and calcined in air at 873 K for six hours [32]. $Cu/MgO/Al_2O_3$ catalysts in a molar ratio of Cu:Mg:Al = 50:30:20 were prepared by the co-precipitation method and were calcined in air at 573 K [34] and 873 K [33] for four hours. The promoted perovskite-type catalysts (La-Cu-Zn-O) were prepared by sol-gel method and calcined in air at 673 K for two hours and then at 1073 K for four hours [35]. MgO-promoted Cu/TiO_2 catalysts used $Mg(NO_3) \cdot 6\ H_2O$ as magnesium source and were calcined in air at 273 K for four hours [36].

Girod et al. [47] reported in a recent study with steel mill gases and Clariant's MegaMax®800 (thyssenkrupp Steel Europe site in Duisburg, Germany) catalyst the feasibility of methanol synthesis based on H_2-enriched blast furnace gases. However, pronounced catalyst deactivation was observed, highlighting the need for further investigation of trace compounds in the cleaned steel mill gas streams and their possible deactivating effects on the catalyst. Reference tests with various synthetic gas compositions also showed catalyst deactivation within the first 100 h under kinetically controlled reaction conditions. Raising the temperature and thereby changing into thermodynamically controlled reaction conditions resulted in constant methanol equilibrium concentration in the product stream without any indication of catalyst deactivation [47].

Bos et al. [48] investigated the synthesis of methanol by direct CO_2 hydrogenation with a commercial $Cu/ZnO/Al_2O_3$ catalyst (CP-488) from Johnson Matthey in a semi-continuous reactor with two temperature zones, one for the reaction and the second one for the in situ condensation of the products. CO_2 conversion of > 99.5% was reported [48]. From the results of continuous admixture of feed gas and discontinuous removal of product condensate the authors concluded that the carbon-based selectivity loss to CO can be neglected, as—similar to our findings—the CO content remained constant after the starting phase.

Neither internal nor external condensation and recycling has gained satisfactory energy efficiency yet. This offers great potential for further investigation. Based on the semi-continuous reactor concept with in situ condensation a conceptual design for methanol production based on a stand-alone wind power plant, CO_2 capture from air, and renewable H_2 produced by water electrolysis has been proposed [49]. With an estimated methanol price of 800 EUR t^{-1}, this concept has not yet been made economically feasible, but it is potentially viable enough to encourage further investigations.

To gain progress in the usage of CO_2-rich industrial off-gas as feedstock, the design of sophisticated reactor concepts, and the development of easy to prepare and recycle catalysts with sufficient catalytic activity will play an important role in the mitigation of industrial CO_2 emissions. The comparison of literature data with the performance of the Cu/MgO catalyst prepared by our group encourages investigation of separate caustification/sintering of the MgO precursor before mixing with the catalytically active constituent(s).

4. Conclusions

Methanol synthesis from CO_2 with a Cu/MgO catalyst was investigated. The topic has been investigated in an ongoing project to collect data about the interaction of sintering temperature dependent MgO reactivity and the catalytic activity of Cu/MgO catalysts. In a first experimental series of catalyst preparation the caustification temperature of the MgO carrier material was limited to a level of 823 K to obtain highly active MgO with respect to CO_2 adsorption capacity. Then, the MgO carrier was impregnated with copper nitrate, calcined and activated, and then tested in a semi-continuous bench scale tank reactor setup. The results of this first series of experiments confirm the catalytic activity of the catalyst as prepared. The results indicate that the activity of the catalyst, as prepared, still becomes better after 48 h of operation. From the results of this study, it is concluded that in methanol synthesis Cu/MgO catalysts with high caustic reactivity of MgO provide sufficient activity. The results of this study offer a profound basis for further investigation of MgO-based catalysts with different caustic reactivity. To gain results about the role of the carrier material MgO the effect of different caustification/sintering temperature levels on the activity has to be investigated in next steps. However, these investigations will need an improved determination of the caustic reactivity of MgO and complete chemical and morphological analysis of the catalyst to identify the effect of MgO quality on the catalytic activity of Cu/MgO catalysts.

Author Contributions: Conceptualization, S.K. and S.L.; methodology, S.K., S.L. and M.S.; software, S.K.; validation, S.K. and S.L.; formal analysis, S.K.; investigation, S.K. and M.P.; data curation, S.K.; writing—original draft preparation, S.K.; writing—review and editing, S.K., S.L. and M.S.; visualization, S.K.; supervision, S.L.; project administration, S.K. and S.L. All authors have read and agreed to the published version of the manuscript.

Funding: This research received no external funding.

Institutional Review Board Statement: Not applicable.

Informed Consent Statement: Not applicable.

Acknowledgments: The authors gratefully acknowledge the support from the NAWI Graz program. Thanks are also due to Tanja Weiß, Mario Kircher, and Christian Winter for their support with laboratory work. Open Access Funding by the Graz University of Technology.

Conflicts of Interest: The authors declare no conflict of interest.

References

1. Friedlingstein, P.; Jones, M.W.; O'Sullivan, M.; Andrew, R.M.; Hauck, J.; Peters, G.P.; Peters, W.; Pongratz, J.; Sitch, S.; le Quéré, C.; et al. Global Carbon Budget 2019. *Earth Syst. Sci. Data* **2019**, *11*, 1783–1838. [CrossRef]
2. Jiang, X.; Nie, X.; Guo, X.; Song, C.; Chen, J.G. Recent Advances in Carbon Dioxide Hydrogenation to Methanol via Heterogeneous Catalysis. *Chem. Rev.* **2020**, *120*, 7984–8034. [CrossRef]
3. Jadhav, S.G.; Vaidya, P.D.; Bhanage, B.M.; Joshi, J.B. Catalytic carbon dioxide hydrogenation to methanol: A review of recent studies. *Chem. Eng. Res. Des.* **2014**, *92*, 2557–2567. [CrossRef]
4. Din, I.U.; Shaharun, M.S.; Alotaibi, M.A.; Alharthi, A.I.; Naeem, A. Recent developments on heterogeneous catalytic CO_2 reduction to methanol. *J. CO2 Util.* **2019**, *34*, 20–33. [CrossRef]
5. Dang, S.; Yang, H.; Gao, P.; Wang, H.; Li, X.; Wei, W.; Sun, Y. A review of research progress on heterogeneous catalysts for methanol synthesis from carbon dioxide hydrogenation. *Catal. Today* **2019**, *330*, 61–75. [CrossRef]
6. Olah, G.A. Beyond Oil and Gas: The Methanol Economy. *Angew. Chem. Int. Ed.* **2005**, *44*, 2636–2639. [CrossRef] [PubMed]
7. Yang, H.; Zhang, C.; Gao, P.; Wang, H.; Li, X.; Zhong, L.; Wei, W.; Sun, Y. A review of the catalytic hydrogenation of carbon dioxide into value-added hydrocarbons. *Catal. Sci. Technol.* **2017**, *7*, 4580–4598. [CrossRef]
8. Álvarez, A.; Bansode, A.; Urakawa, A.; Bavykina, A.V.; Wezendonk, T.A.; Makkee, M.; Gascon, J.; Kapteijn, F. Challenges in the Greener Production of Formates/Formic Acid, Methanol, and DME by Heterogeneously Catalyzed CO_2 Hydrogenation Processes. *Chem. Rev.* **2017**, *117*, 9804–9838. [CrossRef]
9. Zhao, Y.-F.; Yang, Y.; Mims, C.; Peden, C.H.; Li, J.; Mei, D. Insight into methanol synthesis from CO_2 hydrogenation on Cu(111): Complex reaction network and the effects of H_2O. *J. Catal.* **2011**, *281*, 199–211. [CrossRef]
10. Deutschmann, O.; Knözinger, H.; Kochloefl, K.; Turek, T. Heterogeneous Catalysis and Solid Catalysts, 3. Industrial Applications. In *Ullmann's Encyclopedia of Industrial Chemistry*; Wiley: Chichester, UK, 2010.
11. Gurau, B.; Smotkin, E.S. Methanol crossover in direct methanol fuel cells: A link between power and energy density. *J. Power Sources* **2002**, *112*, 339–352. [CrossRef]
12. Breeze, P.A. *Power Generation Technologies*, 3rd ed.; Newnes: Kidlington, UK; Oxford, UK, 2019.
13. Day, W.H. Methanol Fuel in Commercial Operation on Land and Sea. *Gas Turbine World*. 2016. Available online: https://www.methanol.org/wp-content/uploads/2016/12/Methanol-Nov-Dec-2016-GTW-.pdf (accessed on 7 June 2021).
14. Bertau, M.; Offermanns, H.; Plass, L.; Schmidt, F.; Wernicke, H.-J. *Methanol: The Basic Chemical and Energy Feedstock of the Future*; Springer: Berlin/Heidelberg, Germany, 2014.
15. Ma, J.; Sun, N.; Zhang, X.; Zhao, N.; Xiao, F.; Wei, W.; Sun, Y. A short review of catalysis for CO_2 conversion. *Catal. Today* **2009**, *148*, 221–231. [CrossRef]
16. Ott, J.; Gronemann, V.; Pontzen, F.; Fiedler, E.; Grossmann, G.; Kersebohm, D.B.; Weiss, G.; Witte, C. Methanol. In *Ullmann's Encyclopedia of Industrial Chemistry*; Wiley: Chichester, UK, 2010; pp. 1–27.
17. Outotec Technologies. *HSC Chemistry®Software*; Outotec Technologies: Helsinki, Finland, 2018.
18. Raudaskoski, R.; Turpeinen, E.; Lenkkeri, R.; Pongrácz, E.; Keiski, R.L. Catalytic activation of CO_2: Use of secondary CO_2 for the production of synthesis gas and for methanol synthesis over copper-based zirconia-containing catalysts. *Catal. Today* **2009**, *144*, 318–323. [CrossRef]
19. Arena, F.; Mezzatesta, G.; Zafarana, G.; Trunfio, G.; Frusteri, F.; Spandaro, L. How oxide carriers control the catalytic functionality of the Cu-ZnO system in the hydrogenation of CO_2 to methanol. *Catal. Today* **2013**, *210*, 39–46. [CrossRef]
20. Qi, G.-X.; Zheng, X.-M.; Fei, J.-H.; Hou, Z.-Y. Low-temperature methanol synthesis catalyzed over Cu/γ-Al_2O_3-TiO_2 for CO_2 hydrogenation. *Catal. Lett.* **2001**, *72*, 191–196. [CrossRef]
21. Chinchen, G.C.; Denny, P.J.; Parker, D.G.; Spencer, M.S.; Whan, D.A. Mechanism of methanol synthesis from CO_2/CO/H_2 mixtures over copper/zinc oxide/alumina catalysts: Use of ^{14}C-labelled reactants. *Appl. Catal.* **1987**, *30*, 333–338. [CrossRef]
22. Bowker, M. The mechanism of methanol synthesis on copper/zinc oxide/alumina catalysts. *J. Catal.* **1988**, *109*, 263–273. [CrossRef]

23. Portha, J.-F.; Parkhomenko, K.; Kobl, K.; Roger, A.-C.; Arab, S.; Commenge, J.-M.; Falk, L. Kinetics of Methanol Synthesis from Carbon Dioxide Hydrogenation over Copper-Zinc Oxide Catalysts. *Ind. Eng. Chem. Res.* **2017**, *56*, 13133–13145. [CrossRef]
24. Aresta, M.; Dibenedetto, A. Utilisation of CO_2 as a chemical feedstock: Opportunities and challenges. *Dalton Trans.* **2007**, 2975–2992. [CrossRef] [PubMed]
25. Razali, N.A.M.; Lee, K.T.; Bhatia, S.; Mohamed, A.R. Heterogeneous catalysts for production of chemicals using carbon dioxide as raw material: A review. *Renew. Sustain. Energy Rev.* **2012**, *16*, 4951–4964. [CrossRef]
26. Agny, R.M.; Takoudis, C.G. Synthesis of methanol from carbon monoxide and hydrogen over a copper-zinc oxide-alumina catalyst. *Ind. Eng. Chem. Prod. Res. Dev.* **1985**, *24*, 50–55. [CrossRef]
27. Henrici-Olivé, G.; Olivé, S. Mechanistic reflections on the methanol synthesis with Cu/Zn catalysts. *J. Mol. Catal.* **1982**, *17*, 89–92. [CrossRef]
28. Fujitani, T.; Saito, M.; Kanai, Y.; Kakumoto, T.; Watanabe, T.; Nakamura, J.; Uchijima, T. The role of metal oxides in promoting a copper catalyst for methanol synthesis. *Catal. Lett.* **1994**, *25*, 271–276. [CrossRef]
29. Robbins, J.L.; Iglesia, E.; Kelkar, C.P.; DeRites, B. Methanol synthesis over Cu/SiO_2 catalysts. *Catal. Lett.* **1991**, *10*, 1–10. [CrossRef]
30. Baltes, C.; Vukojevic, S.; Schuth, F. Correlations between synthesis, precursor, and catalyst structure and activity of a large set of $CuO/ZnO/Al_2O_3$ catalysts for methanol synthesis. *J. Catal.* **2008**, *258*, 334–344. [CrossRef]
31. Nielsen, N.D.; Thrane, J.; Jensen, A.D.; Christensen, J.M. Bifunctional Synergy in CO Hydrogenation to Methanol with Supported Cu. *Catal. Lett.* **2020**, *150*, 1427–1433. [CrossRef]
32. Ren, H.; Xu, C.-H.; Zhao, H.-Y.; Wang, Y.-X.; Liu, J.; Liu, J.-Y. Methanol synthesis from CO_2 hydrogenation over Cu/γ-Al_2O_3 catalysts modified by ZnO, ZrO_2 and MgO. *J. Ind. Eng. Chem.* **2015**, *28*, 261–267. [CrossRef]
33. Dasireddy, V.D.; Neja, S.Š.; Likozar, B. Correlation between synthesis pH, structure and $Cu/MgO/Al_2O_3$ eterogeneous catalyst activity and selectivity in CO_2 hydrogenation to methanol. *J. CO2 Util.* **2018**, *28*, 189–199. [CrossRef]
34. Dasireddy, V.D.; Štefančič, N.S.; Huš, M.; Likozar, B. Effect of alkaline earth metal oxide (MO) $Cu/MO/Al_2O_3$ catalysts on methanol synthesis activity and selectivity via CO_2 reduction. *Fuel* **2018**, *233*, 103–112. [CrossRef]
35. Zhan, H.; Li, F.; Gao, P.; Zhao, N.; Xiao, F.; Wei, W.; Zhong, L.; Sun, Y. Methanol synthesis from CO_2 hydrogenation over La–M–Cu–Zn–O (M = Y, Ce, Mg, Zr) catalysts derived from perovskite-type precursors. *J. Power Sources* **2014**, *251*, 113–121. [CrossRef]
36. Liu, C.; Guo, X.; Guo, Q.; Mao, D.; Yu, J.; Lu, G. Methanol synthesis from CO_2 hydrogenation over copper catalysts supported on MgO-modified TiO_2. *J. Mol. Catal. A Chem.* **2016**, *425*, 86–93. [CrossRef]
37. Zander, S.; Kunkes, E.L.; Schuster, M.E.; Schumann, J.; Weinberg, G.; Teschner, D.; Jacobsen, N.; Schlögl, R.; Behrens, M. The role of the oxide component in the development of copper composite catalysts for methanol synthesis. *Angew. Chem. Int. Ed.* **2013**, *52*, 6536–6540. [CrossRef]
38. Yang, R.; Zhang, Y.; Iwama, Y.; Tsubaki, N. Mechanistic study of a new low-temperature methanol synthesis on Cu/MgO catalysts. *Appl. Catal. A Gen.* **2005**, *288*, 126–133. [CrossRef]
39. Kim, H.Y.; Lee, H.M.; Park, J.-N. Bifunctional Mechanism of CO_2 Methanation on Pd-MgO/SiO_2 Catalyst: Independent Roles of MgO and Pd on CO_2 Methanation. *J. Phys. Chem. C* **2010**, *114*, 7128–7131. [CrossRef]
40. Loder, A.; Siebenhofer, M.; Lux, S. The reaction kinetics of CO_2 methanation on a bifunctional Ni/MgO catalyst. *J. Ind. Eng. Chem.* **2020**, *85*, 196–207. [CrossRef]
41. Seeger, M.; Otto, W.; Flick, W.; Bickelhaupt, F.; Akkerman, O.S. Magnesium Compounds. In *Ullmann's Encyclopedia of Industrial Chemistry*; Wiley: Chichester, UK, 2010.
42. Jin, F.; Al-Tabbaa, A. Thermogravimetric study on the hydration of reactive magnesia and silica mixture at room temperature. *Thermochim. Acta* **2013**, *566*, 162–168. [CrossRef]
43. Yang, N.; Ning, P.; Li, K.; Wang, J. MgO-based adsorbent achieved from magnesite for CO_2 capture in simulate wet flue gas. *J. Taiwan Inst. Chem. Eng.* **2018**, *86*, 73–80. [CrossRef]
44. Salomão, R.; Pandolfelli, V.C. Magnesia sinter hydration–dehydration behavior in refractory castables. *Ceram. Int.* **2008**, *34*, 1829–1834. [CrossRef]
45. Toth, A.; Schnedl, S.; Painer, D.; Siebenhofer, M.; Lux, S. Interfacial Catalysis in Biphasic Carboxylic Acid Esterification with a Nickel-Based Metallosurfactant. *ACS Sustain. Chem. Eng.* **2019**, *7*, 18547–18553. [CrossRef]
46. Dasireddy, V.D.; Likozar, B. The role of copper oxidation state in $Cu/ZnO/Al_2O_3$ catalysts in CO_2 hydrogenation and methanol productivity. *Renew. Energy* **2019**, *140*, 452–460. [CrossRef]
47. Girod, K.; Lohmann, H.; Kaluza, S. Methanol Synthesis with Steel Mill Gases: Performance Investigations in an On-Site Technical Center. *Chem. Ing. Tech.* **2021**, *93*, 850–855. [CrossRef]
48. Bos, M.J.; Brilman, D. A novel condensation reactor for efficient CO_2 to methanol conversion for storage of renewable electric energy. *Chem. Eng. J.* **2015**, *278*, 527–532. [CrossRef]
49. Bos, M.J.; Kersten, S.; Brilman, D. Wind power to methanol: Renewable methanol production using electricity, electrolysis of water and CO_2 air capture. *Appl. Energy* **2020**, *264*, 114672. [CrossRef]

Article

Integration of Renewable Hydrogen Production in Steelworks Off-Gases for the Synthesis of Methanol and Methane

Michael Bampaou [1,2,*], Kyriakos Panopoulos [2,*], Panos Seferlis [1,2], Spyridon Voutetakis [1], Ismael Matino [3], Alice Petrucciani [3], Antonella Zaccara [3], Valentina Colla [3], Stefano Dettori [3], Teresa Annunziata Branca [3] and Vincenzo Iannino [3]

[1] Centre for Research and Technology Hellas (CERTH), Chemical Process and Energy Resources Institute (CPERI), 57001 Thessaloniki, Greece; seferlis@auth.gr (P.S.); paris@certh.gr (S.V.)
[2] Department of Mechanical Engineering, Aristotle University of Thessaloniki, 54124 Thessaloniki, Greece
[3] TeCIP Institute, Scuola Superiore Sant'Anna, Via Moruzzi 1, 56124 Pisa, Italy; i.matino@santannapisa.it (I.M.); alice.petrucciani@santannapisa.it (A.P.); antonella.zaccara@santannapisa.it (A.Z.); valentina.colla@santannapisa.it (V.C.); s.dettori@santannapisa.it (S.D.); teresa.branca@santannapisa.it (T.A.B.); v.iannino@santannapisa.it (V.I.)
* Correspondence: bampaou@certh.gr (M.B.); panopoulos@certh.gr (K.P.); Tel.: +30-2310498286 (M.B.)

Abstract: The steel industry is among the highest carbon-emitting industrial sectors. Since the steel production process is already exhaustively optimized, alternative routes are sought in order to increase carbon efficiency and reduce these emissions. During steel production, three main carbon-containing off-gases are generated: blast furnace gas, coke oven gas and basic oxygen furnace gas. In the present work, the addition of renewable hydrogen by electrolysis to those steelworks off-gases is studied for the production of methane and methanol. Different case scenarios are investigated using AspenPlus^{TM} flowsheet simulations, which differ on the end-product, the feedstock flowrates and on the production of power. Each case study is evaluated in terms of hydrogen and electrolysis requirements, carbon conversion, hydrogen consumption, and product yields. The findings of this study showed that the electrolysis requirements surpass the energy content of the steelwork's feedstock. However, for the methanol synthesis cases, substantial improvements can be achieved if recycling a significant amount of the residual hydrogen.

Keywords: blast furnace gas; coke oven gas; basic oxygen furnace gas; methanation; methanol synthesis; aspen plus; gas cleaning; hydrogen; steelworks sustainability

1. Introduction

The iron and steel industry is among the industrial sectors with the highest production volumes, having indispensable end-products for modern society [1]. The European steel industry, in particular, is a world leader in steel production accounting for approximately 16% of the world production (8.5% belongs to the European Union countries), coming second only to China. In market and economic terms, in 2019 it generated 140 bn € of gross added value and employed around 2.67 million people [2]. Steelworks, however, are one the most energy- and carbon-intensive industries in the world, accounting for 27% of the total industrial CO_2 emissions and 4–5% of the total anthropogenic CO_2 emissions [3]. Since world steel production is expected to rise in the following years, CO_2 and carbon emissions will increase accordingly, if no proper countermeasures are adopted [1].

During the primary steel production route, carbonaceous off-gases are generated during the main production steps of: (1) conversion of coal to coke in the coke oven, (2) pig iron production in the blast furnace and (3) processing of pig iron to steel in the basic oxygen furnace [1]. Since the usage of fossil fuels (usually coal and natural gas) as reducing agents in the blast furnace is intensely optimized [4], alternative ways are investigated for the reduction in those emissions. Generally, a common way to avoid the flaring of steelmaking off-gases is their use as internal energy sources both for heating and power production. As a consequence, a reduction in natural gas and external produced electricity use is obtained, resulting in a decrease in emissions, primary resources consumption, and operating costs. Recent works focused on the optimization of the management of steelworks off-gases networks using a decision support system [5] including machine learning-based forecasting models [6–8] and advanced optimization strategies [9,10].

An alternative/complementary way of utilizing those gases without deviating from the already established steel production route is their conversion to added-value chemicals. The proposed utilization strategy involves the use of the carbonaceous feedstocks for the production of methane and methanol (MeOH) through the addition of renewable hydrogen by proton exchange membrane (PEM) electrolysis. Apart from the environmental perspective, target is to partially replace the fossil fuel demands of the steel plant and/or to generate revenue by utilizing a by-product stream. Methanol has already broad commercial uses, as chemical intermediate and fuel [11], whereas methane apart from its commercial value, can be used within the steel plant for power production and/or reused as reducing agent in the BF process [12]. The proposed strategy, however, has to surpass or match the benefits obtained through the conventional off-gases exploitation strategy (i.e., heating and power production) from an energetic, economic and environmental perspective.

The three mentioned steelworks off-gases (Blast Furnace Gas: BFG, Coke Oven Gas: COG, Basic Oxygen Furnace Gas: BOFG) are commonly stored in dedicated gasholders, that act as buffers. These off-gases contain more or less the same compounds but at different proportions: the most common are CO_2, CO, H_2, and N_2. Small amounts of impurities are also contained; however, they do not pose environmental threats when combusted in their traditional use within the plant. However, when advanced catalytic processes are pursued using these gases as feedstock, then, further gas cleaning steps are required to avoid catalyst poisoning. The present work considers an already existing gas cleaning setup prior to the gas holder short storage as the starting point for the process formulation. In addition, further gas cleaning steps are proposed upstream the catalytic processes, considering a possible presence of residual impurities in the off-gases, before entering the catalytic syntheses units.

The scope of this work is to study the integration of renewable hydrogen into steelworks off-gases for the efficient production of methane and methanol and to exploit the largest amounts of steelworks off-gases as carbon sources. This is a novelty in respect to past works [13,14] that exploit only limited amounts of these off-gases and focus mainly on the exploitation of the COG as feedstock for the synthesis reactors (as it is or mixed with other off-gases, due to its high hydrogen content). This study has been conducted using flowsheet simulations in AspenPlus^{TM}. The key points of this work can be summarized as follows:

- Based on the possible contained impurities, a gas cleaning strategy is proposed in order to avoid poisoning of the synthesis catalysts.
- The modelling methodology both for methane and methanol synthesis is presented, and sensitivity analyses are conducted to define specific operating parameters with a major influence on the overall process.
- Five case studies are analyzed, which correspond to different utilization amounts of the steelworks off-gases for the production of methane/methanol, whereas one of the investigated cases involves the combination of methane and methanol production. The defined case studies also include the discussion and analysis of PEM electrolysis for renewable hydrogen production.
- The overall benefits of these scenarios are compared to the traditional use of power production in energetic and efficiency terms. Process improvements are proposed to increase the overall efficiency of the integrated process.

This work is organized as follows: Section 2 describes the main features and characteristics of the considered off-gases; Section 3 illustrates the investigated process and the comprising sub-systems; and Section 4 presents and discusses the obtained results. Finally, Section 5 provides the conclusions of this work and hints for future work.

2. Steelworks Off-Gases

Table 1 depicts the total volumetric amounts and the mean composition of the steelworks off-gases, for a steel plant producing 6 MT steel per year [15].

Table 1. Mean composition of steelworks off-gases.

Mean Composition (mol.%)	BFG	COG	BOFG
CO	23.5	4.1	54.0
Ar + O_2	0.6	0.2	0.7
H_2	3.7	60.7	3.2
CO_2	21.6	1.2	20.0
N_2	46.6	5.8	18.1
CH_4	0.0	22.0	0.0
$\Sigma C_n H_m$	0.0	2.0	0.0
H_2O	4.0	4.0	4.0
Potential impurities ([1,12,16])	\multicolumn{3}{c}{H_2S, SO_2, Organic sulfur, HCN, NH_3, NO_x, BTX, halogens, heavy metals}
Total amount (m^3/h)	730,000	40,000	35,000

As shown in Table 1 from an overall perspective, the component that prevails through the three gases is nitrogen. The inert nature of nitrogen lowers the partial pressures of the reactants and raises the volume of the feed gases. Thus, it increases the capital expenses and the costs associated to compression and could lead to accumulation within a recycling loop. The contained CO and CO_2 can be used as feedstock for the production of chemicals (e.g., CH_4 and/or CH_3OH). The reactivity of CO is always higher than the one of CO_2 for a considered chemical, resulting, thus, in higher activation energies for the CO_2 conversion [15]. The insufficient amount of H_2 contained in the off-gases, dictates the addition of additional hydrogen for the synthesis. In order to increase the carbon efficiency of the process, attention should be paid on the addition of renewable hydrogen instead of fossil based. In addition, it is assumed that after cleaning, the off-gases are water saturated. This water content should be removed prior to compression, in order to: (i) avoid condensation that could damage the compressors [15], and (ii) avoid the promotion of the Water Gas Shift (WGS) reaction that could lead to the consumption of CO for the formation of additional CO_2 [17]. Finally, the off-gases also contain small amounts of oxygen, which need to be removed for safety reasons prior to the conduction of adsorption processes, such as pressure swing adsorption [18].

From a particular point of view, the BFG contains large quantities of nitrogen due to the use of hot air as oxidant within the furnace and low amount of H_2 [6]; enrichment is required for its use as feedstock for methane and methanol production. The coke oven gas is generated in the coking plant during the heating of coal to produce coke. In contrast to the BFG, it contains large amounts of hydrogen and can be mixed with the other gases to reduce the required amounts of additional hydrogen by electrolysis. In addition, it can be easily valorized within the plant as fuel or feedstock for producing chemicals, due to the contained hydrogen and methane [19]. The BOFG is generated in the basic oxygen furnace, where oxygen is injected to oxidize part of the carbon in the pig iron produced from the blast furnace; it contains predominantly CO [15].

The three steelworks off-gases are generally used internally for heating and electricity production purposes. For instance, the COG is used for firing coke ovens, as heat input for rolling mills and to produce energy at the power plant [20]. The blast furnace gas serves also as a fuel for firing the coke ovens, the hot blast stoves heating the wind to be injected into the blast furnace and the power plant [21], whereas the basic oxygen furnace gas, apart from power applications, can also be used for upgrading the heating value of the BFG in a gas mixing station [16]. Gasholders are used for storing the surplus of those gases. However, in some cases the gasholders capacities are not sufficient to contain the generated quantities and as a consequence, the excess off-gases are flared. In other cases, the gases do not satisfy the internal requirements and natural gas is purchased. An optimized off-gas distribution management can improve the efficiency [5], as well as the consideration of an alternative use, such as for methane and methanol production. Regarding the available amounts for CH_4 and MeOH production, it is assumed that 50% of the total generated amount is available, after the rest being utilized in internal applications within the plant [16,20].

3. Process Description

In this section, the outline of the five case studies is described. These case studies differ on the quantities of the utilized gases for the syntheses and on the produced chemical (methane/methanol). The case studies are:

- 100% utilization of the available by-product gases (BFG, COG, and BOFG) for the production of methane.
- Methanation of 80% of the available by-product gases and the remaining fraction is used in the power plant.
- Methanation of specific amounts of the by-product gases in order to replace the natural-gas demands of the plant and the remaining fraction is used in the power plant.
- Methanol synthesis of 80% of the available by-product gases and the remaining fraction is used in the power plant.
- Methanation of specific amounts of the by-product gases in order to replace the natural-gas demands of the plant and 50% of the quantity used in Case 4 is used for the production of significant quantities of MeOH (only the remaining by-product gases are used in the power plant).

The selected scenarios want to cover the short-, medium-, and long-term technology deployment horizon and to provide useful information in order to reduce the relevant costs when moving towards the large deployment of the proposed technological option. In particular, the Cases 1, 2, and 4 are adopted because they actually present the medium- to long-term capacities required to deploy the proposed technological option. On the other hand, Cases 3 and 5 represent a shorter-term demonstration of that technological option to move towards decarbonization of steelmaking.

After the description of the case studies in Section 3.1, the overall process scheme is presented in Section 3.2 that includes the aforementioned systems (gas cleaning, hydrogen production, methane, and methanol synthesis) as well as the power plant. A gas cleaning strategy is proposed in Section 3.3 based on the possible contained impurities, whereas Sections 3.4 and 3.5 involve the description of the methanol and methane synthesis processes and the followed AspenPlus^{TM} modelling methodology. Sensitivity analyses on crucial modelling approaches and operating parameters are conducted on both processes. Finally, Section 3.6 includes the description of PEM electrolysis for the production of renewable hydrogen.

3.1. Case Studies Description

The integration of methane and methanol synthesis is evaluated for the previously described scenarios considering a steelmaking plant of medium size with an annual steel production of about 6 MT. The different case scenarios are evaluated in terms of carbon conversion, product yields, hydrogen requirements and consumption, electrolysis demands, as well as overall efficiency of the process. For the cases where power is produced, it is assumed that a gas-fired boiler is in operation within the plant [16]. Figures 1–5 depict the different flowrates and utilization factors of the steelworks off-gases.

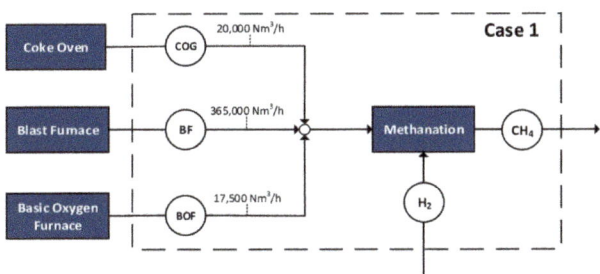

Figure 1. Case 1—100% of off-gases as input for methanation.

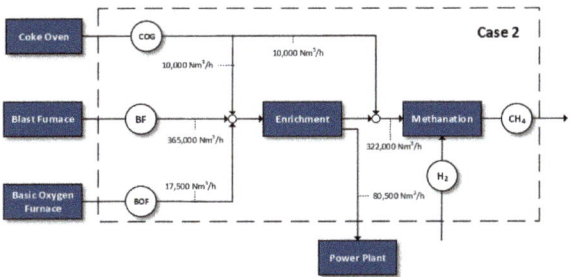

Figure 2. Case 2—methanation of 80% of by-product gases.

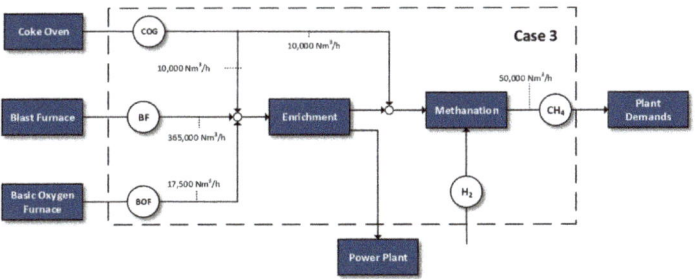

Figure 3. Case 3—replacement of natural gas demands by methanation.

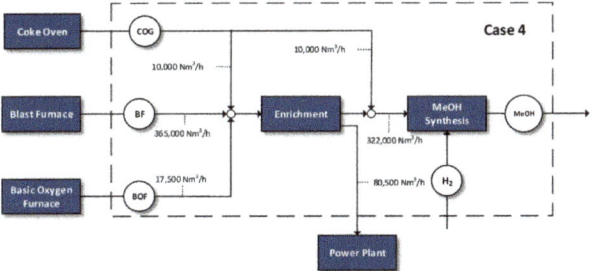

Figure 4. Case 4—methanol synthesis of 80% of by-product gases.

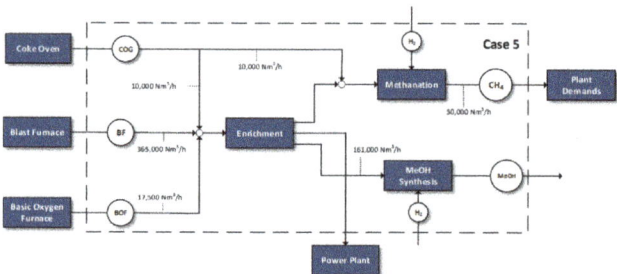

Figure 5. Case 5—100% replacement of natural gas demands and methanol synthesis.

1. 100% utilization of the produced by-product gases for the production of methane

The first case represents the utilization of the entire available amount of the three off-gases for the production of methane. Renewable hydrogen is added in basic stoichiometric ratio to produce methane. This represents a boundary scenario for the utilization of the steelwork gases, which is restrictive in terms of hydrogen flows and electrolysis power requirements.

2. Methanation of 80% of the available by-product gases and the remaining fraction used in the power plant

The steelworks off-gases have generally various uses within the steel plant and the main one is for the production of electrical power. However, in this scenario, 80% of the total amounts of the gases are used for the production of methane with the addition of renewable hydrogen and the other fraction is sent to the power plant. Before the methanation process, the enrichment step serves as mixing/upgrading process before entering the power plant. A part of the COG is dispatched directly to methane synthesis, due to its higher hydrogen content compared to the other gases. Although, the CH_4 content of COG could have a negative impact in methanation activity, the amount of available COG is relatively small compared to the total utilized gases, reducing significantly the methane quantities (i.e., <1%) in the reactor inlet of the methanation cases (i.e., after the H_2 addition).

3. Methanation of specific amounts of the by-product gases in order to replace the natural gas demands of the plant

This case investigates the possibility of valorizing the steelworks off-gases for the replacement of the internal steelworks needs of natural gas—assuming an overestimated case of approximately 50,000 Nm^3/h internal natural gas demands for a 6 MT/year steel plant [12]. The remaining portion of the gases is combusted in the power plant.

4. Methanol synthesis of 80% of the by-product gases and the rest goes to the power plant

Similar with Case 2, 80% of the amounts of the gases are used for methanol production with the addition of renewable hydrogen, whereas the remaining portion is used in the power plant.

5. Methanation of specific amounts of the by-product gases in order to replace the natural gas demands and the production of significant quantities of MeOH

Case 5 represents the most integrated valorization scheme for steel gases for the simultaneous production of methane and methanol. After the enrichment step, half of the amount employed in Case 4 is used for methanol production and another part is used for the replacement of the industry's natural gas demands. Finally, a part of the gases is sent to the power plant.

Figure 6 shows a comparison of the energy content of the mixed feedstock of Case 1 with respect to the energy content of the other cases, by highlighting the different off-gases contribution. In the first scenario, BFG comprises 73% of the total energy content of the feed stream to the syntheses processes, due to the larger used flowrate, whereas COG, although in lower quantity (20,000 Nm^3/h compared to 365,000 Nm^3/h of the BFG; 5% of the total amount) contains a significant portion of the overall energy content (19%). This indicates a higher energy content per m^3, due to the contained CH$_4$ and H$_2$ and the lower CO$_2$ and N$_2$ contents in the COG feedstock. Regarding the energy contents of the feedstock used in the different scenarios, Cases 2 and 4 use the same feed quantities for the production of chemicals (81% of the energy compared to Case 1). Case 3 represents the feedstock energy content for the replacement of the natural gas demand of the plant, while Case 5 is a combined case for the replacement of natural gas as well as for methanol synthesis (33% and 70%, respectively).

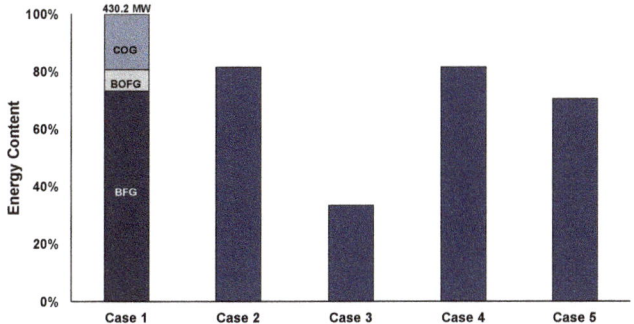

Figure 6. Feedstock energy content for the different case scenarios.

3.2. Integration Options of CH$_4$/MeOH Syntheses Concepts into Steelworks

Figure 7 shows the overall process flowsheet that includes the major sections of the proposed concept: gas conditioning, methane production, methanol synthesis, hydrogen production, and the power plant (for the cases where power is produced).

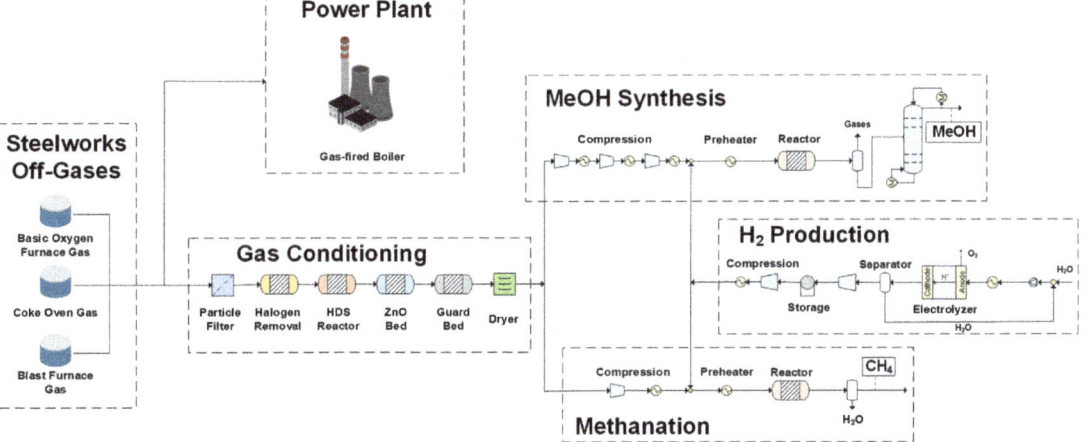

Figure 7. Integration of synthesis units into steelworks—superstructure flowsheet.

The mixture of the steelworks off-gases, after an ad hoc conditioning for removal of unwanted impurities, is fed either to the methanol synthesis or to the methanation section. For the methanol synthesis, the feed gas undergoes compression in three stages in order to reduce the associated compression ratio costs; intermediate cooling between the stages is provided. Afterwards, hydrogen is added to reach the required stoichiometric number and the inlet mixture is preheated before inserted to the synthesis reactor. The produced mixture is separated using a flash separator into the liquid (mainly methanol and water) and gaseous products that consist of the unreacted hydrogen and the rest of the initial feedstock. The last step is the purification of the methanol product in a distillation column, which removes the contained product water.

The first step for methanation requires compression only in one step, since methanation takes place at low pressures (<10 bar). Renewable hydrogen is added to achieve the required stoichiometric ratio and the inlet feed is preheated and directed to the reactor. A flash separator is also used to separate the gaseous products from the produced liquid water. Table 2 depicts the assumptions that refer to the overall process flowsheet simulations. The off-gases composition reported in Table 1 is taken as the starting point of the subsequent flowsheet simulations.

Table 2. General assumptions and specifications of the overall process.

	General Assumptions	Reference
Property method	Soave–Redlich–Kwong equation of state (SRK)	[22,23]
Input feed temperature	25 °C	Assumption
Input feed pressure	1 bar	Assumption
Compression stages	Methanol: 3, methane: 1	Assumption
Input H_2 purity, vol.%	H_2: 99.9%, H_2O: 0.1%	[24]
Power plant efficiency	Gas-fired boiler: 40%	[16]
H_2 consumption	$\frac{[H_2]_{in} - [H_2]_{out}}{[H_2]_{in}}$	-
CO conversion	$\frac{[CO]_{in} - [CO]_{out}}{[CO]_{in}}$	-
CO_2 conversion	$\frac{[CO_2]_{in} - [CO_2]_{out}}{[CO_2]_{in}}$	-
Carbon conversion	$\frac{([CO]_{in} + [CO_2]_{in}) - ([CO]_{out} + [CO_2]_{out})}{[CO]_{in} + [CO_2]_{in}}$	-
Methanol yield	$\frac{[CH_3OH]_{out}}{[CO]_{in} + [CO_2]_{in}}$	-
CH_4 yield	$\frac{[CH_4]_{out} - [CH_4]_{in}}{[CO]_{in} + [CO_2]_{in}}$	-

3.3. Impurities and Gas Conditioning

The three steelwork off-gases undergo different cleaning steps in order to remove the contained, unwanted components before being stored in gas holders. Typical gas cleaning steps involve dust removal, cooling, scrubbing (for ammonia and BTX removal), and demistering [19,25]. After the initial steps, additional gas cleaning is required to protect the methane [26] and methanol [27] syntheses catalysts.

As shown in Table 1, several sulfur-containing compounds can be found in the steelworks off-gases, which cause corrosion and poisoning of Cu-based catalysts. Other common impurities include nitrogen-containing species such as ammonia or hydrogen cyanide. At the high temperatures of the steel production processes, nitrogen oxides NO_x can be formed, which have to be removed from the exhaust gases, whereas at lower temperatures, NH_3 can be adsorbed at catalyst sites, reducing the catalyst activity [28]. Halogens (HCl, HF, and HBr) are also contained in the off-gases and are known to cause corrosion and poison catalysts. In particular, experimental works have shown that HCl poisoning could cause loss of the active surface area of the catalyst and promote sintering of the copper crystallites [27]. Furthermore, additional reactions could occur between HCl and other contaminant-forming species such as NH_4Cl and NaCl, which when condensed, could cause fouling and create deposits in cooler downstream pipes and equipment [29,30]. Finally, trace elements and heavy metals are also contained in the off-gases due to the diverse nature of the feedstocks. Besides corrosion problems, other trace elements pose a threat to human health and the environment. The distribution and partitioning of these contaminants play an important role on the undertaken cleaning strategy. For example, particle filters could be used for solid particles, but if those compounds appear in the gaseous phase, more advanced cleaning efforts should be employed, such as solid sorption. Whether a trace element appears in the gas or particulate phase and in which form, depends on following factors [31]:

- how the trace element resides in the incoming material,
- temperature and pressure,
- oxidizing or reducing conditions,
- presence of halogens, such as chlorine,
- presence of compounds that can act as sorbents, such as calcium.

Based on the contained impurities, Figure 8 depicts the proposed off-gases cleaning strategy. Each of the cleaning steps targets aims at a specific impurity group. However, possible interactions between an impurity and a precedent/succeeding step cannot be ruled out.

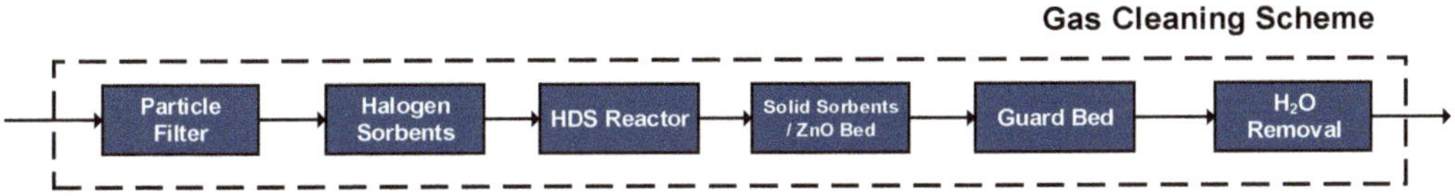

Gas Cleaning Scheme

Particle Filter → Halogen Sorbents → HDS Reactor → Solid Sorbents / ZnO Bed → Guard Bed → H_2O Removal

Figure 8. Proposed gas cleaning scheme.

A first step is devoted to the removal of any contained solid particles through fine filters. Afterwards, the contained halogens (HCl, HF, etc.) are removed using inexpensive sorption materials such as $NaHCO_3$ (Nahcolite) or Trona (Na_2CO_3-$NaHCO_3$-$2H_2O$) [28]. For instance, in the case of nahcolite, HCl is removed in the form of NaCl, whereas H_2O and CO_2 are also formed, according to the following reaction:

$$NaHCO_3 + HCl \rightarrow NaCl + H_2O + CO_2.$$

Regarding the sulfur-containing compounds, H_2S is more easily removed at ppb levels with respect to other sulfur species. A common strategy consists in converting organic sulfur compounds to H_2S and then employing adsorption technologies for the deep removal of H_2S [32]. The avoidance of acid gas removal process, such as Selexol^{TM} or Rectisol^{TM}, despite their efficiency in reducing H_2S to ppm levels, lies within their affinity to physically absorb CO_2, which should otherwise be used as feedstock for the production of methanol/methane [26].

At the hydrodesulfurization (HDS) reactor, organic sulfur compounds and COS are converted to H_2S through the addition of hydrogen. The usual employed catalysts are based on cobalt and nickel. A possible reaction network for the conversion to hydrogen sulfide is the following [33]:

$$COS + H_2 \rightarrow H_2S + CO$$

$$CS_2 + 4H_2 \rightarrow 2H_2S + CH_4.$$

Afterwards, a sorption bed containing metal oxides, such as CaO and ZnO, can be used for the removal of H_2S. For the case of ZnO, H_2S is removed in the form of ZnS [34]:

$$H_2S + ZnO \rightarrow H_2O + ZnS.$$

It is an exothermic process, conducted at T < 250 °C and as shown in the reaction stoichiometry, the reaction equilibrium is not affected by pressure, whereas the inlet content of water could affect the H_2S removal efficiency. Studies have shown that H_2S can be effectively removed at ppb levels employing the ZnO strategy [34,35]. However, due to the contained CO and CO_2, additional reactions could occur, with a consequent deterioration of the H_2S removal efficiency [34,36].

Finally, a guard bed is placed, containing nickel or other inexpensive material to protect the subsequent synthesis units. It restricts impurities that could have escaped from the former gas cleaning steps and acts as a final protection before the production of chemicals. In addition, the gases are dried to remove the contained water to avoid condensation during compression and/or the promotion of unwanted side-reactions.

3.4. Methanol Synthesis

3.4.1. Process Description and Modelling Approach

Methanol synthesis is based on the following three reactions:

$$CO + 2H_2 \rightarrow CH_3OH \quad \Delta H^0 = -90 \text{ kJ/mol}_{CO}$$

$$CO_2 + 3H_2 \rightarrow CH_3OH + H_2O \quad \Delta H^0 = -49 \text{ kJ/mol}_{CO2}$$

$$CO_2 + H_2 \rightarrow CO + H_2O \quad \Delta H^0 = +41 \text{ kJ/mol}_{CO2}.$$

The first two hydrogenation reactions can be combined to form the reverse water gas shift (RWGS) reaction, indicating thus a dependency in-between the reaction system [37]. The catalytic methanol synthesis is exothermic and thermodynamically favored by lower temperatures and higher pressures. Today most of the world methanol production is covered with natural gas derived synthesis gas that after H_2/CO ratio adjustment is catalytically processed at 50–100 bar and temperatures between 200–300 °C (temperatures required for the activation of the employed catalyst) [38]. An alternative consideration could be a process occurring at much higher pressures (above 100 bar), which would result into an increase in the CO and CO_2 conversion rates and thus lowering the needs for carbon recycling [39,40]. This would, however, result in increasing compression costs and power demands and therefore, it was not adopted in this study.

The most common MeOH catalyst employed in industrial scale is based on $CuO/ZnO/Al_2O_3$, which is also considered in this study. At higher synthesis temperatures, sintering could take place resulting in higher deactivation rate of the catalyst [41]. The produced water, mainly by CO_2 hydrogenation, apart from affecting the equilibrium, could also adsorb on the catalyst sites and promote catalyst sintering [42]. In past works, in-situ water removal was proposed for the enhancement of the thermodynamic equilibrium concentration [43]. The methanol synthesis reaction is characterized by the stoichiometric number (S.N.) where $[H_2]$, $[CO]$, and $[CO_2]$ refer to the molar flows of the feed components: S.N. $= \frac{[H_2]-[CO_2]}{[CO]+[CO_2]}$. A value of S.N. = 2 refers to a stoichiometric correlation between the components, whereas the optimum case is slightly above the stoichiometric number [41].

Methanol synthesis applications result in conversion close to what the thermodynamic equilibrium dictates. Any additional hydrogen is not consumed throughout the process and remains unexploited [38]. Therefore, the process economics could benefit from separating the residual H_2 and reuse it in the synthesis reactor.

In this work, MeOH synthesis reactor is simulated using two different approaches: a thermodynamic and a kinetic approach. The thermodynamic approach is represented by an AspenPlus^{TM} RGibbs reactor model (based on Gibbs free energy minimization), which for a given pressure and temperature, calculates the equilibrium concentration of selected components. The kinetic approach utilizes the kinetic model developed by Vanden Bussche (with WHSV = 2 kg$_{feed}$ kg$_{cat}^{-1}$ h^{-1} and Bed Voidage: 0.33) [44]. Simulation results have shown that for the studied conditions, the deviation of the two modelling approaches is within an acceptable range (<5%) and therefore, the thermodynamic approach is being employed in the investigations.

3.4.2. Modelling Assumptions

The modelling of methanol synthesis is based on chemical equilibrium by means of minimization of the Gibbs free energy. Certain components included in the feed mixture are assumed as inert components, having thus no influence in the reaction. Apart from nitrogen, ethane, and methane are also treated as inert gases that do not affect reaction equilibrium. The property method that is used in the flowsheet simulations is Soave–Redlich–Kwong equation of state [45], as past works have proven that it is suitable for methanol synthesis applications [46,47]. Table 3 shows the assumptions and specifications for the thermodynamic MeOH synthesis model.

Table 3. General assumptions and specifications of the MeOH AspenPlus^{TM} model.

	General Assumptions
Property method	Soave–Redlich–Kwong equation of state (SRK)
Feed preheating	150 °C
Reaction temperature	200–300 °C
Reaction pressure	50–100 bar
Reactor type	Thermodynamic—RGibbs
Possible products	H_2, CO, CO_2, H_2O, CH_3OH, C (solid), O_2
Inert components	N_2, CH_4, C_2H_6
H_2 stoichiometry/Stoichiometric number (S.N.)	$\frac{[H_2]-[CO_2]}{[CO]+[CO_2]} = 1.7\text{--}2.1$

3.4.3. Sensitivity Analysis

Since the methanol synthesis reaction is exothermic, it is thermodynamically favored by lower temperatures. However, the temperature range of the catalyst's activation should also be taken into consideration in order to find the optimum operating conditions. Higher pressures are thermodynamically preferred for methanol production (Figure 9a–c), but the higher compression costs should also be taken into account. Figure 9c depicts the lower conversion rate of CO_2 compared to CO, whereas at higher temperature and pressure values, CO conversion tends to decrease and CO_2 to increase. This fact can be attributed to the WGS reaction, which is an endothermic reaction and is thermodynamically favored by higher temperatures.

Figure 9d–f shows the influence of increasing stoichiometric number, e.g., increasing input hydrogen. Higher stoichiometric numbers result in higher methanol yields for a given operating temperature (Figure 9d). However, relatively to the input, less hydrogen is consumed in the reactor (Figure 9e) and more remains unexploited in the outlet gaseous fraction (Figure 9f), which refers to the gaseous stream after the separation of methanol and water. Figure 9g illustrates the need for drying of the feed mixture before entering the synthesis reactor. It can be seen that an increase in the water content of the inlet feed leads to a strong decrease in the maximum attained methanol yield. This behavior can be attributed to the promotion of the WGS reaction and consequent CO conversion to additional CO_2, at the expense of the methanol synthesis reactions.

The higher the input hydrogen flowrates, the higher the quantity that remains unexploited during the process. Even in sub-stoichiometric ratios, the remaining hydrogen is in considerable portions, which illustrates the need for efficient hydrogen management throughout the process. This could be achieved either through operating in sub-stoichiometric numbers or employing hydrogen recirculation technologies to lower the needs for additional hydrogen and increase the overall efficiency of the system.

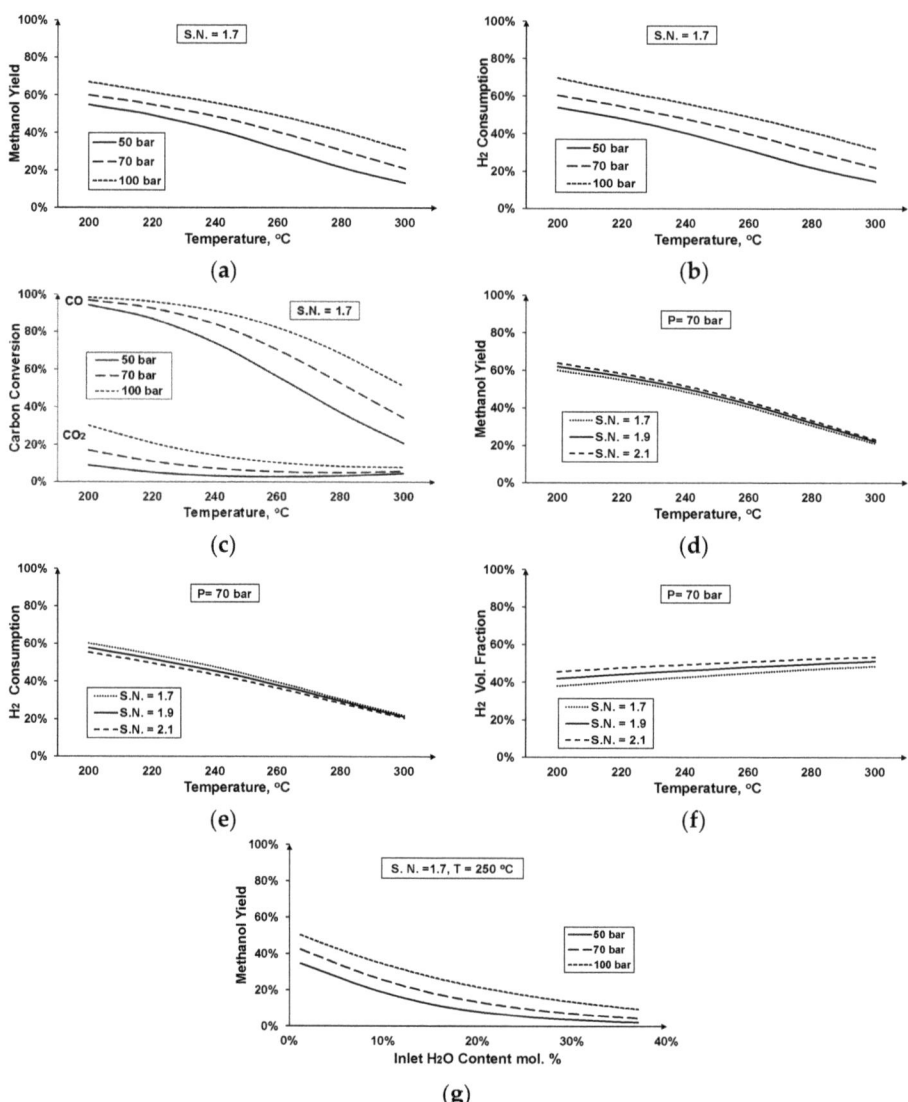

Figure 9. Methanol synthesis at different operating conditions: (**a**) methanol yield at constant stoichiometric number, (**b**) hydrogen consumption at constant stoichiometric number, (**c**) carbon conversion at constant stoichiometric number, (**d**) methanol yield at constant pressure, (**e**) hydrogen consumption at constant pressure, (**f**) residual hydrogen at constant pressure, and (**g**) methanol yield—inlet water content at constant stoichiometric number and temperature.

3.5. Methane Production

3.5.1. Process Description and Modelling Approach

Syngas methanation is a highly exothermic process aiming at the production of Substitute Natural Gas (SNG) from CO and CO_2 with the addition of H_2 at the required stoichiometries. The simplicity and high efficiency of the process have been crucial parameters for the establishment of this technology for the production of methane from waste feedstocks such as biomass [48,49] or steelwork off-gases [19]. For the production of methane from syngas, the main occurring reactions are:

$$CO + 3H_2 \rightarrow CH_4 + H_2O \qquad \Delta H^0 = -206 \text{ kJ/mol}_{CO}$$

$$CO_2 + 4H_2 \rightarrow CH_4 + 2H_2O \qquad \Delta H^0 = -165 \text{ kJ/mol}_{CO2}$$

$$CO_2 + H_2 \rightarrow CO + H_2O \qquad \Delta H^0 = +41 \text{ kJ/mol}_{CO2}.$$

However, based on experimental results, it is assumed to consist of a more complex reaction network, taking provision also for the formation of solid carbon throughout the process [22].

A variety of catalysts are employed for the catalytic methanation process. In this work, a nickel-based catalyst is considered, since it is mostly employed in commercial applications due to the high activity and low associated costs [50]. Similar to methanol synthesis, methane production is also favored by low temperatures and higher pressures because it results in reduction in the total volume. There is currently a variety of established methanation concepts operating at different conditions and reactor configurations [50]. In this work, methanation is conducted in low temperatures 200–300 °C and pressures < 10 bar.

Again, two approaches based on kinetics and thermodynamics are compared using the reactor inlet composition of Case 1, as a comparison basis. The first approach is based on the Langmuir-Hinshelwood (LHHW) kinetics derived from Kopyscinski et al. (for WHSV = 2×10^{-5} kg$_{feed}$ kg$_{cat}^{-1}$ h^{-1} and Bed Voidage = 0.33) [51]. Because the LHHW kinetics cannot be implemented directly in AspenPlus^{TM}, a revised form was adopted [52]. The second approach is also based on the thermodynamic model using the Gibbs free energy minimization method, as already explained for the methanol synthesis system. Simulation results have shown that the two modelling approaches agree well for the studied conditions (deviation < 5%) and therefore, for the subsequent sensitivity analyses and case studies evaluation, the thermodynamic model was used.

3.5.2. Modelling Assumptions

Modelling of the methanation section is also based on the minimization of the Gibbs free energy. In this case, nitrogen and ethane are treated as inert gases and the used property method in the flowsheet simulations is Soave–Redlich–Kwong equation of state [22,23]. Table 4 illustrates the assumptions for the AspenPlus^{TM} model of methane synthesis.

Table 4. General assumptions and specifications of the AspenPlus^{TM} methanation model.

	General Assumptions
Property method	Soave–Redlich–Kwong equation of state (SRK) [22,23]
Reactor type	Thermodynamic—RGibbs
Possible products	H_2, CO, CO_2, H_2O, CH_3OH, C (solid), O_2, CH_4
Inert components	N_2, C_2H_6
Feed preheating	150 °C
Reaction temperature	200–300 °C
Reaction pressure	1–10 bar
H_2 stoichiometry	$\frac{[H_2]}{3[CO]+4[CO_2]} = 1\text{--}1.1$

3.5.3. Sensitivity Analysis

Figure 10 shows some crucial characteristics of the process at different operating parameters of methane production. As depicted in Figure 10, CO conversion is almost complete, irrespectively of the operating pressure and temperature range (T = 200–300 °C, P = 1–10 bar). On the other hand, CO_2 conversion, strongly depends on the operating parameters; lower temperature and higher pressure favor the CO_2 conversion thermodynamically. Higher conversion rates mean higher hydrogen consumption (see Figure 10b), resulting in >95% consumption in any pressure and temperature range. This results in low portion of the hydrogen remaining unexploited in the off-gases of the methanation process. In addition, Figure 10c shows that methane production is favored in any of these operating conditions obtaining a methane yield greater than 95%.

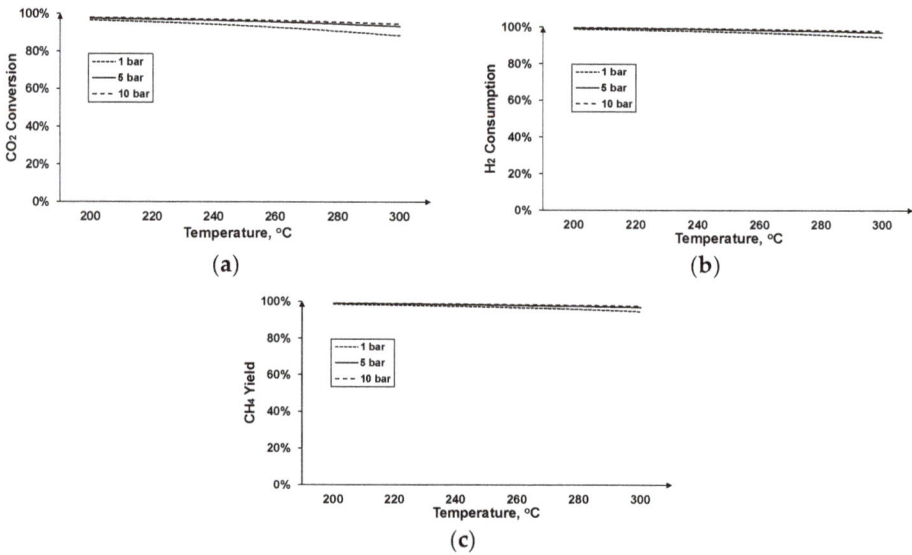

Figure 10. Methane production at different operating temperature and pressure: (**a**) CO_2 conversion, (**b**) hydrogen consumption, and (**c**) methane yield.

3.6. Hydrogen Production

Due to the composition of the steelworks off-gases, hydrogen addition is required in order to reach certain stoichiometric ratios and improve the efficiency of methane and methanol syntheses. In order to obtain both economic and environmental advantages, hydrogen needs to be produced in an environmentally friendly way, i.e., by exploiting renewable sources. In this work, the adopted process is water electrolysis fed by renewable energy.

During water electrolysis, the water molecules are split into hydrogen and oxygen by means of electricity. There are three main electrolysis processes, each one differing on the operating principles and conditions: alkaline exchange membrane (AEM), proton exchange membrane (PEM) and solid oxide electrolysis (SOE) [53]. In the present work, PEM electrolysis is considered as the option for renewable hydrogen production, since it is an already established technology, it is used in large-scale industrial applications and it is not sensitive to the fluctuations in power supply, such as in the case of renewable energy sources [24].

For the calculation of the power requirements, an AspenPlus^{TM} PEM electrolysis model has been developed. The model incorporates the following main reactions occurring in the two PEM sections:

$$2H_2O \rightarrow O_2 + 4H^+ + 2e^- \qquad \text{anode section}$$

$$2H^+ + 2e^- \rightarrow H_2 \qquad \text{cathode section.}$$

In addition, further phenomena taking place inside a PEM electrolysis module are also considered such as hydrogen and oxygen permeations [54,55] and water diffusion [55,56], which are estimated in ad hoc configured calculator blocks. Highly pure hydrogen is assumed to be produced (<99.9 vol.%), and the overall electrical energy consumption for the stack is 54.8 kWh/kg H_2. The produced hydrogen is stored into pressurized vessels and compressed to achieve the conditions required for the syntheses units.

A detailed description of the PEM electrolysis model, as well as of other renewable hydrogen production technologies that are considered possible solutions for the enrichment of steelworks off-gases, is included in another publication by the authors [57].

4. Results

In this section, the proposed case studies are evaluated using the aforementioned AspenPlus^{TM} models. The results presented in Table 5 focus on the hydrogen requirements and consumption, electrolysis demands, product yields, and carbon conversion. Figure 11 shows these key results. Further specific indicative stream results are available in the Appendix A.

For the three first cases, which refer only to methanation, carbon is almost completely converted, compared to Case 4, which includes only methanol synthesis, and Case 5, which is a combination between methane and methanol syntheses. In Case 4, in particular, the low CO_2 utilization rates indicate that CO (see Figure 11a) and not CO_2 (see Figure 11b) is consumed for methanol synthesis. In addition, for the different case studies, the higher the carbon conversion rate, the higher the hydrogen consumption throughout the process (see Figure 11c). Figure 11d shows the produced electrical power of the different cases compared to the base-case, that refers to the traditional, full-scale utilization of the steelworks off-gases for power production. Case 1 is not included in the comparison, as the whole amount of off-gases is used for methane production. For Cases 2 and 4, the same power is produced since the same feedstock amount is used in the syntheses units (19% of the total power). Case 3 produces 60% of the power produced in the base-case and Case 5, which is the most integrated scheme, involves the production of 30% of the total power.

If different stoichiometric ratios are chosen (other than the stoichiometric for methane and 1.7 for methanol synthesis) the required hydrogen feed inputs are greatly affected, as illustrated in Figure 12. Case 1, which refers to the full-scale utilization of the steelworks by-product gases, requires more hydrogen compared to the other cases at any stoichiometric ratio. Although Cases 2 and 4 refer to utilization of the same feedstock flowrates, methane synthesis (Case 2) requires more hydrogen compared to methanol synthesis (Case 5), due to the higher carbon conversion. As a consequence of the lower carbon conversion, the rest of the hydrogen remains unexploited in the off-gases of the methanol synthesis process.

Table 5. Case studies key results (methanation: T = 250 °C, P = 5 bar and stoichiometric H_2, MeOH synthesis: T = 250 °C, P = 70 bar, and S.N. = 1.7).

	Case 1	Case 2	Case 3	Case 4	Case 5
Feed (kg/s)	133.7	106.5	35.9	106.5	90.3
Feed compression (MW)	29.5	23.6	8.3	68.3	42.4
H_2 feed (kg/s)	13.7	10.9	3.5	6.6	7.0
H_2 consumption (%)	98.7	98.7	98.7	36.1	67.9
Carbon conversion (%)	98.2	98.2	98.2	37.0	61.0
CO_2 utilization (%)	96.1	96.1	96.0	3.1	39.3
CH_4 product (kg/s)	32.9	26.3	9.0	-	9.0
CH_3OH product (kg/s)	-	-	-	19.2	9.9
Power production (MW)	-	31.9	103.3	31.9	50.8

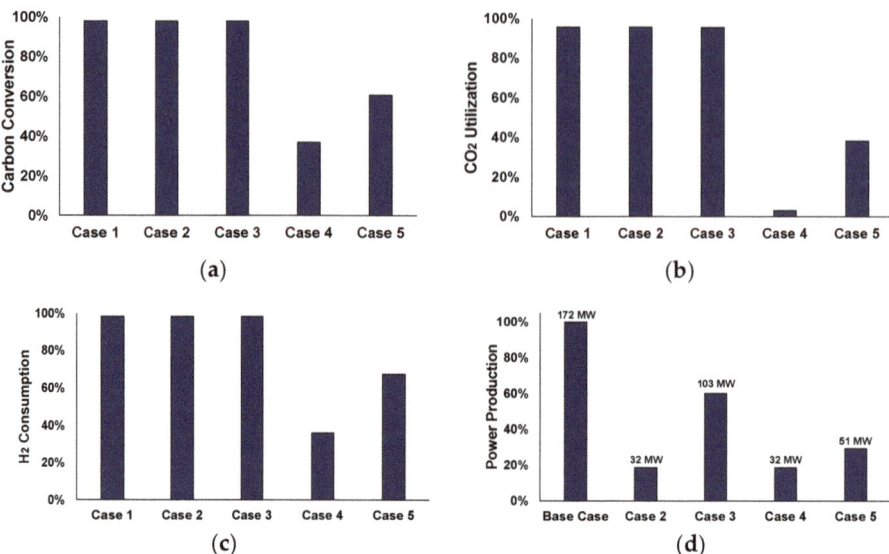

Figure 11. Key Results of the case studies (methanation: T = 250 °C, P = 5 bar and stoichiometric H_2, MeOH synthesis: T = 250 °C, P = 70 bar and S.N. = 1.7): (**a**) carbon conversion, (**b**) CO_2 utilization, (**c**) hydrogen consumption, and (**d**) power production.

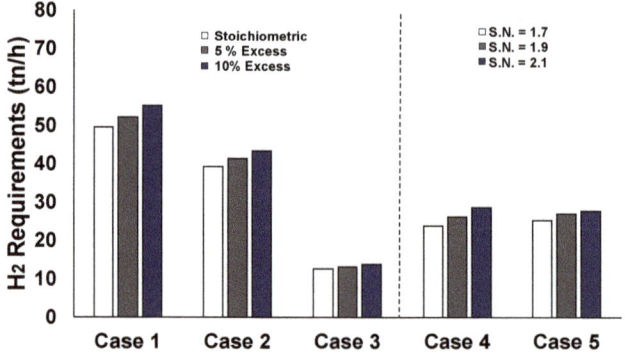

Figure 12. Hydrogen requirements for the different stoichiometries per case.

This is also verified in Figure 13, which depicts the energy content of the residual off-gases after each synthesis processes and after the separation of the total amounts of produced methane and methanol, in each respective case. Especially for Cases 4 and 5, the remaining off-gases have a significant energetic value due to the large fraction of unreacted hydrogen. These residual off-gases could either be used for combustion to support heat-intensive processes, or hydrogen recycling should be included to avoid producing additional hydrogen by electrolysis. In conventional methanol synthesis loops, a flash drum separates the methanol and water products from the unreacted gaseous components, which are recycled back to the synthesis reactor [41]. Alternative processes to recover only the residual hydrogen include technologies such as pressure swing adsorption [58], membranes [59], and/or electrochemical hydrogen compression [60].

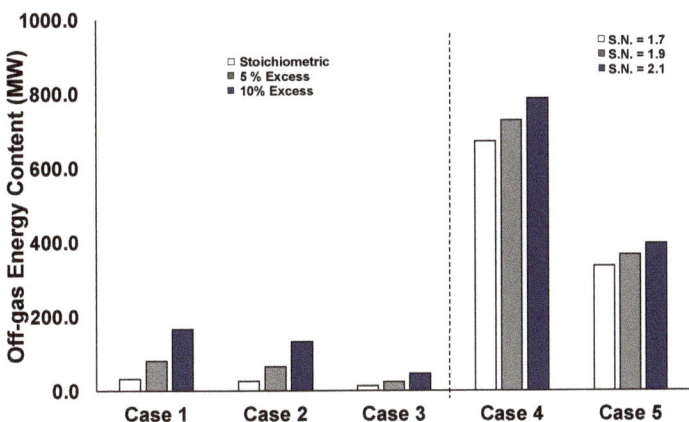

Figure 13. Residual off-gases energy content for the different stoichiometries per case.

The PEM electrolysis requirements, as illustrated in Figure 14, are in the range of GWs, which are restrictive for employment in full-scale, by considering the capacities of currently available commercial electrolyzers.

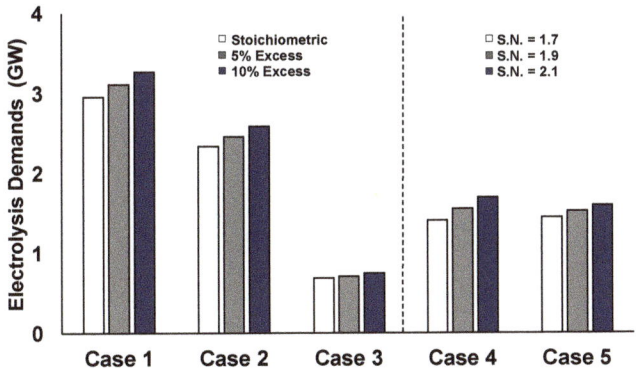

Figure 14. PEM electrolysis requirements for the different cases.

In terms of comparison between the energy content of the total feedstock (used in each case) and the electrolysis requirements, Figure 15 shows the major energy streams of the Cases 1, 4, and 5. In Case 1, 383% of the energy content is contained in the methane product, 89% is contained in the methanol of Case 4, whereas 105% and 46% are contained in the methanol and methanol products of Case 5. The electrolysis requirements of each case are noticeable. In Case 1, 631% of the energy of the feedstock is required for electrolysis, in comparison to 304% and 322% in Cases 4 and 5. Regarding the MeOH synthesis cases, however, these figures could be further reduced, if certain amounts of the residual hydrogen are recycled.

Figure 16 compares the PEM electrolysis energy requirements of the base case without H_2 recycling, to recycling 25%, 50%, and 75% of the residual hydrogen. If recycling 75% of the residual hydrogen is pursued, almost 50% less power is required for electrolysis, indicating, thus opportunity for further optimization of the process.

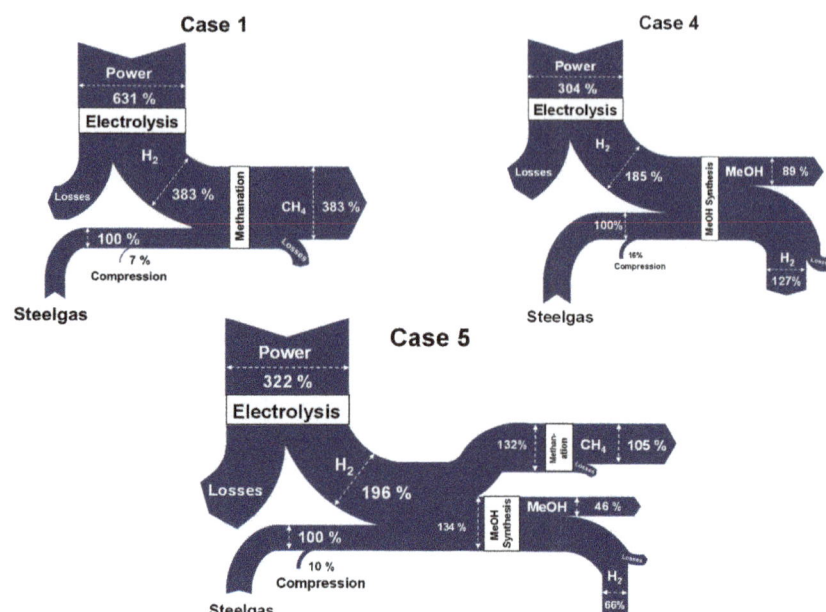

Figure 15. Sankey diagrams—energy analysis for three key scenarios.

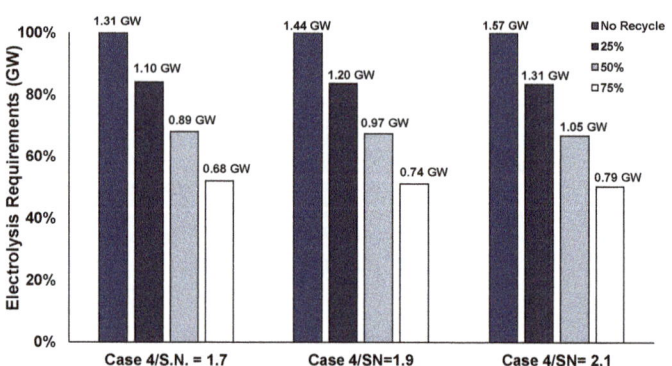

Figure 16. Influence of hydrogen recycling in PEM electrolysis requirements of Case 4.

Regarding Case 5, the benefits of recycling 75% could result in almost 25% lower electrolysis requirements (see Figure 17). The lower savings percentage compared to Case 4 is due to the methanation section of Case 5, which consumes a major fraction of the input hydrogen, resulting in less available hydrogen for recycling, which is also verified in Figure 11.

The benefits of recycling can also be seen in Figure 18, which refer to Cases 4 and 5 and the required electrolysis power to the energy content of the steelworks feedstock (as shown in Figure 15). The 304% of Case 4 could be reduced to 158% and from 322% to 244% for Case 5 if recycling 75% of the hydrogen is pursued.

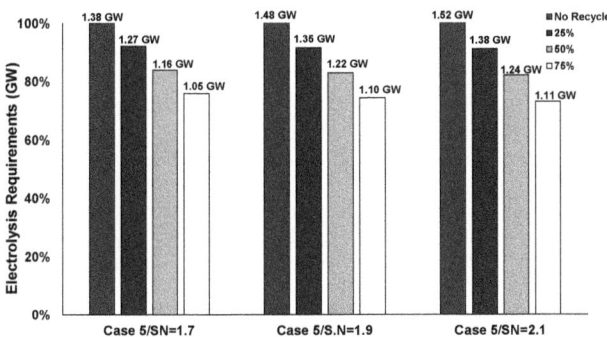

Figure 17. Influence of hydrogen recycling in PEM electrolysis requirements of Case 5.

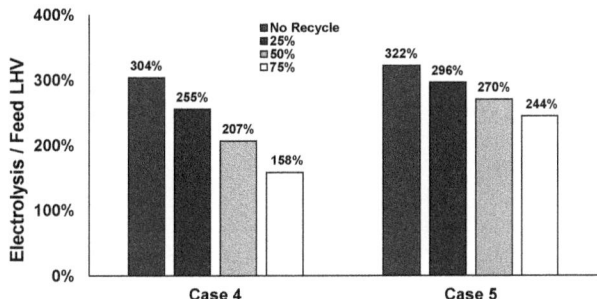

Figure 18. Electrolysis requirements compared to feedstock energy content for the different recycling cases for Case 4 and Case 5.

5. Discussion

In this work, the integration of hydrogen-intensified methane and methanol synthesis is investigated for three available steelworks off-gases by means of AspenPlus^{TM} flowsheet simulations. The composition of the off-gases is analyzed and a generic gas cleaning scheme is proposed for the removal of the contained impurities that can affect the catalyst operation, which is used in the CH_4 and CH_3OH syntheses. Thermodynamic and kinetic AspenPlus^{TM} models are compared for the investigation of methanol and methane synthesis processes. The studied conditions for methane synthesis include 200–300 °C, stoichiometric number 1–1.1 and pressure 1–10 bar and for methanol synthesis: 200–300 °C, stoichiometric number 1.7–2.1, and pressure 50–100 bar. Furthermore, case studies corresponding to different usage of the steelworks off-gases for chemicals production are investigated in terms of hydrogen requirements and consumption, carbon conversion, product yields, and PEM electrolysis requirements. The cases of methane synthesis depict high CO and CO_2 conversion rates that almost eliminate the CO_2 emissions of the steel plant. In case of increasing carbon credits, this would represent a significant financial benefit and therefore, a carbon credit avoidance could be of high importance. On the other hand, methanol synthesis produces a product with higher market value, but only converts approximately 40% of the carbon emissions into a renewable fuel/chemical. The choice between methane and methanol production or a combination of the two, will be a result of an upcoming cost estimation study but also on the relevant prices of the products, as well as, on the carbon credits. The energy content of the hydrogen employed for this transformation far overcomes the energy off-set of the steel plants. However, recycling of the residual hydrogen in the methanol-involved cases could lead to substantial benefits in terms of electrolysis requirements. Calculations show that reductions in the range of 50% for Case 4 and 25% for Case 5 of electrolysis requirements could be achieved, when recycling 75% of the residual hydrogen. Future studies will involve the capital and operating cost estimation analysis of the case studies as well as the application of state-of-the art and alternative hydrogen recirculation technologies for recycling the residual hydrogen from the methanol synthesis cases. In addition, the application of an advanced dispatch controller will be investigated in order to optimize the management of steelworks off-gases among internal users, power plant and syntheses processes, and the related hydrogen requirements.

Author Contributions: Conceptualization, M.B., K.P., I.M. and V.C.; methodology, M.B., K.P., I.M., V.I. and V.C.; software, M.B., K.P. and S.V.; validation, M.B., K.P., I.M. and S.D.; formal analysis, K.P., T.A.B. and V.C.; investigation, M.B. and K.P.; resources, K.P., S.V., P.S. and V.C.; data curation, M.B., K.P., A.P. and A.Z.; writing—original draft preparation, M.B. and K.P.; writing—review and editing, M.B., K.P., I.M., V.C., S.D., V.I., A.P., A.Z., T.A.B., P.S. and S.V.; visualization, M.B. and K.P.; supervision, K.P., P.S. and S.V.; project administration, K.P. and V.C.; funding acquisition, K.P. and V.C. All authors have read and agreed to the published version of the manuscript.

Funding: This research was funded by the European Union through the Research Fund for Coal and Steel (RFCS), Grant Agreement No. 800659.

Institutional Review Board Statement: Not applicable.

Informed Consent Statement: Not applicable.

Data Availability Statement: Not applicable.

Acknowledgments: The work described in this paper was developed within the project entitled "Integrated and intelligent upgrade of carbon sources through hydrogen addition for the steel industry" (i3upgrade, GA No. 800659), which has received funding from the Research Fund for Coal and Steel of the European Union. The sole responsibility of the issues treated in this paper lies with the authors; the Commission is not responsible for any use that may be made of the information contained therein.

Conflicts of Interest: The authors declare no conflict of interest.

Appendix A

Indicative process stream results are presented in the following tables, namely, the steelworks off-gases feedstock, the added hydrogen, the reactor inlet, and outlet of the methanol and methane sections (see Figure 7). Note that the depicted cases refer to methane synthesis conducted at T = 250 °C, P = 5 bar, and stoichiometric H_2, whereas for methanol synthesis, T = 250 °C, P = 70 bar, and S.N. = 1.7.

Table A1. Stream results of Case 1.

Stream Name	Feed	H_2 Feed	Reactor Inlet	Reactor Outlet
Stream number	1	2	3	4
m, kg/s	133.68	13.75	144.11	144.11
T, °C	25	150	150	250
P, bar	1	5	5	5
Mole fraction, %				
H_2	6.51	99.9	63.16	1.32
N_2	43.34	-	17.81	27.86
CO_2	20.52	-	8.43	0.52
CO	23.86	-	9.81	Trace
CH_4	1.09	-	0.45	28.71
O_2	0.58	-	0.24	Trace
H_2O	4.00	0.10	0.06	41.53
C_2H_6	0.10	-	0.04	0.06
CH_3OH	-	-	-	-

Table A2. Stream results of Case 2.

Stream Name	Feed	H$_2$ Feed	Reactor Inlet	Reactor Outlet
Stream number	1	2	3	4
m, kg/s	106.48	10.91	114.74	114.74
T, °C	25	150	150	250
P, bar	1	5	5	5
Mole fraction, %				
H$_2$	6.85	99.9	63.11	1.32
N$_2$	43.1	-	17.80	27.83
CO$_2$	20.40	-	8.42	0.52
CO	23.74	-	9.81	Trace
CH$_4$	1.22	-	0.51	28.78
O$_2$	0.58	-	0.24	Trace
H$_2$O	4.00	0.1	0.06	41.49
C$_2$H$_6$	0.11	-	0.05	0.07
CH$_3$OH	-	-	-	-

Table A3. Stream results of Case 3.

Stream Name	Feed	H$_2$ Feed	Reactor Inlet	Reactor Outlet
Stream number	1	2	3	4
m, kg/s	35.89	3.54	38.50	38.50
T, °C	25	150	150	250
P, bar	1	5	5	5
Mole fraction, %				
H$_2$	10.03	99.90	62.71	1.31
N$_2$	40.90	-	17.71	27.60
CO$_2$	19.26	-	8.34	0.52
CO	22.57	-	9.78	Trace
CH$_4$	2.45	-	1.06	29.37
O$_2$	0.56	-	0.24	Trace
H$_2$O	4.00	0.10	0.06	41.05
C$_2$H$_6$	0.22	-	0.10	0.15
CH$_3$OH	-	-	-	-

Table A4. Stream results of Case 4.

Stream Name	Feed	H$_2$ Feed	Reactor Inlet	Reactor Outlet
Stream number	1	2	3	4
m, kg/s	106.48	6.62	114.77	114.77
T, °C	25	150	150	250
P, bar	1	70	70	70
Mole fraction, %				
H$_2$	6.85	99.90	51.67	40.25
N$_2$	43.1	-	23.34	28.47
CO$_2$	20.40	-	11.04	13.06
CO	23.74	-	12.85	5.30
CH$_4$	1.22	-	0.66	0.81
O$_2$	0.58	-	0.32	trace
H$_2$O	4.00	0.10	0.05	1.24
C$_2$H$_6$	0.11	-	0.06	0.07
CH$_3$OH	-	-	-	10.80

Table A5. Stream results of Case 5—MeOH synthesis.

Stream Name	Feed	H$_2$ Feed	Reactor Inlet	Reactor Outlet
Stream number	1	2	3	4
m, kg/s	54.39	3.48	56.54	56.54
T, °C	25	150	150	250
P, bar	1	70	70	4
Mole fraction, %				
H$_2$	5.13	99.90	51.90	40.55
N$_2$	44.29	-	23.42	28.58
CO$_2$	21.01	-	11.11	13.17
CO	24.37	-	12.88	5.29
CH$_4$	0.56	-	0.30	0.36
O$_2$	0.59	-	0.31	trace
H$_2$O	4.00	0.10	0.05	1.21
C$_2$H$_6$	0.05	-	0.03	0.03
CH$_3$OH	-	-	-	10.81

Table A6. Stream results of Case 5—methane production.

Stream Name	Feed	H$_2$ Feed	Reactor Inlet	Reactor Outlet
Stream number	1	2	3	4
m, kg/s	35.89	3.54	38.50	38.50
T, °C	25	150	150	250
P, bar	1	5	5	5
Mole fraction, %				
H$_2$	10.03	99.90	62.71	1.31
N$_2$	40.90	-	17.71	27.60
CO$_2$	19.26	-	8.34	0.52
CO	22.57	-	9.78	trace
CH$_4$	2.45	-	1.06	29.37
O$_2$	0.56	-	0.24	trace
H$_2$O	4.00	0.10	0.06	41.05
C$_2$H$_6$	0.22	-	0.1	0.15
CH$_3$OH	-	-	-	-

References

1. Ramírez-Santos, Á.A.; Castel, C.; Favre, E. A review of gas separation technologies within emission reduction programs in the iron and steel sector: Current application and development perspectives. *Sep. Purif. Technol.* **2018**, *194*, 425–442. [CrossRef]
2. The European Steel Association (EUROFER). *European Steel in Figures 2020*; The European Steel Association (EUROFER): Brussels, Belgium, 2020.
3. Quader, M.A.; Ahmed, S.; Ghazilla, R.A.R.; Ahmed, S.; Dahari, M. A comprehensive review on energy efficient CO2 breakthrough technologies for sustainable green iron and steel manufacturing. *Renew. Sustain. Energy Rev.* **2015**, *50*, 594–614. [CrossRef]
4. Frey, A.; Goeke, V.; Voss, C. Steel Gases as Ancient and Modern Challenging Resource; Historical Review, Description of the Present, and a Daring Vision. *Chem. Ing. Tech.* **2018**, *90*, 1384–1391. [CrossRef]
5. Colla, V.; Matino, I.; Dettori, S.; Petrucciani, A.; Zaccara, A.; Weber, V.; Salame, S.; Zapata, N.; Bastida, S.; Wolff, A.; et al. Assessing the efficiency of the off-gas network management in integrated steelworks. *Mater. Tech.* **2019**. [CrossRef]
6. Matino, I.; Dettori, S.; Colla, V.; Weber, V.; Salame, S. Forecasting blast furnace gas production and demand through echo state neural network-based models: Pave the way to off-gas optimized management. *Appl. Energy* **2019**. [CrossRef]
7. Colla, V.; Matino, I.; Dettori, S.; Cateni, S.; Matino, R. Reservoir Computing Approaches Applied to Energy Management in Industry. In Proceedings of the International Conference on Engineering Applications of Neural Networks, Xersonisos, Greece, 24–26 May 2019; pp. 66–79.
8. Dettori, S.; Matino, I.; Colla, V.; Speets, R. Deep Echo State Networks in Industrial Applications. In Proceedings of the IFIP International Conference on Artificial Intelligence Applications and Innovations, Neos Marmaras, Greece, 5–7 June 2020; pp. 53–63.
9. Maddaloni, A.; Matino, R.; Matino, I.; Dettori, S.; Zaccara, A.; Colla, V. A quadratic programming model for the optimization of off-gas networks in integrated steelworks. *Mater. Tech.* **2019**, *107*. [CrossRef]

20. Wolff, A.; Mintus, F.; Bialek, S.; Dettori, S.; Colla, V. Economical Mixed-Integer Model Predictive Controller for optimizing the sub-network of the BOF gas. In Proceedings of the METEC 4th ESTAD, Düsseldorf, Germany, 24–28 June 2019.
21. Leonzio, G.; Zondervan, E.; Foscolo, P.U. Methanol production by CO$_2$ hydrogenation: Analysis and simulation of reactor performance. *Int. J. Hydrogen Energy* **2019**. [CrossRef]
22. Bender, M.; Roussiere, T.; Schelling, H.; Schuster, S.; Schwab, E. Coupled Production of Steel and Chemicals. *Chem. Ing. Tech.* **2018**, *90*, 1782–1805. [CrossRef]
23. Kim, S.; Kim, J. The optimal carbon and hydrogen balance for methanol production from coke oven gas and Linz-Donawitz gas: Process development and techno-economic analysis. *Fuel* **2020**. [CrossRef]
24. Man, Y.; Yang, S.; Qian, Y. Integrated process for synthetic natural gas production from coal and coke-oven gas with high energy efficiency and low emission. *Energy Convers. Manag.* **2016**. [CrossRef]
25. Uribe-Soto, W.; Portha, J.F.; Commenge, J.M.; Falk, L. A review of thermochemical processes and technologies to use steelworks off-gases. *Renew. Sustain. Energy Rev.* **2017**, *74*, 809–823. [CrossRef]
26. Remus, R.; Roudier, S.; Aguado Monsonet, M.A.; Sancho, L.D. *Best Available Techniques (BAT) Reference Document for Iron and Steel Production*; European Commission: Sevilla, Spain, 2013; ISBN 9789279264757.
27. Saito, M.; Murata, K. Development of high performance Cu/ZnO-based catalysts for methanol synthesis and the water-gas shift reaction. *Catal. Surv. Asia* **2004**, *8*, 285–294. [CrossRef]
28. Wiesmann, T.; Hamel, C.; Kaluza, S. Techniques to Remove Traces of Oxygen by Catalytic Conversion from Gas Mixtures. *Chem. Ing. Tech.* **2018**, *90*, 1446–1452. [CrossRef]
29. Razzaq, R.; Li, C.; Zhang, S. Coke oven gas: Availability, properties, purification, and utilization in China. *Fuel* **2013**, *113*, 287–299. [CrossRef]
30. Kim, Y.-K.; Lee, E.-B. Optimization simulation, using steel plant off-gas for power generation: A life-cycle cost analysis approach. *Energies* **2018**, *11*, 2884. [CrossRef]
31. He, H.; Guan, H.; Zhu, X.; Lee, H. Assessment on the energy flow and carbon emissions of integrated steelmaking plants. *Energy Rep.* **2017**. [CrossRef]
32. Gao, J.; Wang, Y.; Ping, Y.; Hu, D.; Xu, G.; Gu, F.; Su, F. A thermodynamic analysis of methanation reactions of carbon oxides for the production of synthetic natural gas. *RSC Adv.* **2012**. [CrossRef]
33. Koytsoumpa, E.I.; Karellas, S. Equilibrium and kinetic aspects for catalytic methanation focusing on CO2 derived Substitute Natural Gas (SNG). *Renew. Sustain. Energy Rev.* **2018**. [CrossRef]
34. Bertuccioli, L.; Chan, A.; Hart, D.; Lehner, F.; Madden, B.; Standen, E. Development of water electrolysis in the EU. *Fuel Cells Hydrog. Jt. Undert.* **2014**. [CrossRef]
35. Lajtonyi, A. Blast furnace gas cleaning systems. *Millenium Steel* **2006**, *2006*, 57–65.
36. Koytsoumpa, E.I.; Atsonios, K.; Panopoulos, K.D.; Karellas, S.; Kakaras, E.; Karl, J. Modelling and assessment of acid gas removal processes in coal-derived SNG production. *Appl. Therm. Eng.* **2015**, *74*, 128–135. [CrossRef]
37. Spencer, M.S. The role of zinc oxide in Cu/ZnO catalysts for methanol synthesis and the water-gas shift reaction. *Top. Catal.* **1999**, *8*, 259–266. [CrossRef]
38. Rhyner, U. Gas cleaning. In *Synthetic Natural Gas from Coal and Dry Biomass, and Power-to-Gas Applications*; John Wiley & Sons Inc.: Hoboken, NJ, USA, 2016; ISBN 9781119191339.
39. Woolcock, P.J.; Brown, R.C. A review of cleaning technologies for biomass-derived syngas. *Biomass Bioenergy* **2013**, *52*, 54–84. [CrossRef]
40. Ud Din, Z.; Zainal, Z.A. Biomass integrated gasification-SOFC systems: Technology overview. *Renew. Sustain. Energy Rev.* **2016**, *53*, 1356–1376. [CrossRef]
41. Zevenhoven, R.; Kilpinen, P. *Control of Pollutants in Flue Gases and Fuel Gases*; Helsinki University of Technology: Espoo/Turku, Finland, 2001; ISBN 951-22-5527-8.
42. Dou, B.; Wang, C.; Chen, H.; Song, Y.; Xie, B.; Xu, Y.; Tan, C. Research progress of hot gas filtration, desulphurization and HCl removal in coal-derived fuel gas: A review. *Chem. Eng. Res. Des.* **2012**, *90*, 1901–1917. [CrossRef]
43. Dou, B.; Zhang, M.; Gao, J.; Shen, W.; Sha, X. High-temperature removal of NH3, organic sulfur, HCl, and tar component from coal-derived gas. *Ind. Eng. Chem. Res.* **2002**, *41*, 4195–4200. [CrossRef]
44. Li, L.; King, D.L. H2S removal with ZnO during fuel processing for PEM fuel cell applications. *Catal. Today* **2006**, *116*, 537–541. [CrossRef]
45. Novochinskii, I.I.; Song, C.; Ma, X.; Liu, X.; Shore, L.; Lampert, J.; Farrauto, R.J. Low-temperature H2S removal from steam-containing gas mixtures with ZnO for fuel cell application. 1. ZnO particles and extrudates. *Energy Fuels* **2004**, *18*, 576–583. [CrossRef]
46. Sasaoka, E.; Taniguchi, K.; Hirano, S.; Uddin, M.A.; Kasaoka, S.; Sakata, Y. Catalytic Activity of ZnS Formed from Desulfurization Sorbent ZnO for Conversion of COS to H2S. *Ind. Eng. Chem. Res.* **1995**. [CrossRef]
47. Graaf, G.H.; Winkelman, J.G.M.; Stamhuis, E.J.; Beenackers, A.A.C.M. Kinetics of the three phase methanol synthesis. *Chem. Eng. Sci.* **1988**. [CrossRef]
48. Schittkowski, J.; Ruland, H.; Laudenschleger, D.; Girod, K.; Kähler, K.; Kaluza, S.; Muhler, M.; Schlögl, R. Methanol Synthesis from Steel Mill Exhaust Gases: Challenges for the Industrial Cu/ZnO/Al2O3 Catalyst. *Chem. Ing. Tech.* **2018**, *90*, 1419–1429. [CrossRef]

39. Gaikwad, R.; Bansode, A.; Urakawa, A. High-pressure advantages in stoichiometric hydrogenation of carbon dioxide to methanol. *J. Catal.* **2016**. [CrossRef]
40. Bansode, A.; Urakawa, A. Towards full one-pass conversion of carbon dioxide to methanol and methanol-derived products. *J. Catal.* **2014**. [CrossRef]
41. Bozzano, G.; Manenti, F. Efficient methanol synthesis: Perspectives, technologies and optimization strategies. *Prog. Energy Combust. Sci.* **2016**, *56*, 71–105. [CrossRef]
42. Liu, G.; Willcox, D.; Garland, M.; Kung, H.H. The role of CO_2 in methanol synthesis on CuZn oxide: An isotope labeling study. *J. Catal.* **1985**. [CrossRef]
43. Zachopoulos, A.; Heracleous, E. Overcoming the equilibrium barriers of CO_2 hydrogenation to methanol via water sorption: A thermodynamic analysis. *J. CO2 Util.* **2017**. [CrossRef]
44. Vanden Bussche, K.M.; Froment, G.F. A steady-state kinetic model for methanol synthesis and the water gas shift reaction on a commercial Cu/ZnO/Al_2O_3 catalyst. *J. Catal.* **1996**. [CrossRef]
45. Soave, G. Equilibrium constants from a modified Redlich-Kwong equation of state. *Chem. Eng. Sci.* **1972**. [CrossRef]
46. Graaf, G.H.; Sijtsema, P.J.J.M.; Stamhuis, E.J.; Joosten, G.E.H. Chemical equilibria in methanol synthesis. *Chem. Eng. Sci.* **1986**. [CrossRef]
47. Graaf, G.H.; Winkelman, J.G.M. Chemical Equilibria in Methanol Synthesis Including the Water-Gas Shift Reaction: A Critical Reassessment. *Ind. Eng. Chem. Res.* **2016**. [CrossRef]
48. Schildhauer, T.J. Methanation for Synthetic Natural Gas Production—Chemical Reaction Engineering Aspects. In *Synthetic Natural Gas from Coal and Dry Biomass, and Power-to-Gas Applications*; John Wiley & Sons Inc.: Hoboken, NJ, USA, 2016; ISBN 9781119191339.
49. Koytsoumpa, E.I.; Karellas, S.; Kakaras, E. Modelling of Substitute Natural Gas production via combined gasification and power to fuel. *Renew. Energy* **2019**. [CrossRef]
50. Rönsch, S.; Schneider, J.; Matthischke, S.; Schlüter, M.; Götz, M.; Lefebvre, J.; Prabhakaran, P.; Bajohr, S. Review on methanation -From fundamentals to current projects. *Fuel* **2016**, *166*, 276–296. [CrossRef]
51. Kopyscinski, J.; Schildhauer, T.J.; Vogel, F.; Biollaz, S.M.A.; Wokaun, A. Applying spatially resolved concentration and temperature measurements in a catalytic plate reactor for the kinetic study of CO methanation. *J. Catal.* **2010**. [CrossRef]
52. Er-rbib, H.; Bouallou, C. Modeling and simulation of CO methanation process for renewable electricity storage. *Energy* **2014**. [CrossRef]
53. Bessarabov, D.; Wang, H.; Li, H.; Zhao, N. *PEM Electrolysis for Hydrogen Production: Principles and Applications*; CRC Press: Boca Raton, FL, USA, 2015.
54. Barbir, F. PEM electrolysis for production of hydrogen from renewable energy sources. *Sol. Energy* **2005**. [CrossRef]
55. Rivera-Tinoco, R.; Farran, M.; Bouallou, C.; Auprêtre, F.; Valentin, S.; Millet, P.; Ngameni, J.R. Investigation of power-to-methanol processes coupling electrolytic hydrogen production and catalytic CO2 reduction. *Int. J. Hydrogen Energy* **2016**. [CrossRef]
56. Medina, P.; Santarelli, M. Analysis of water transport in a high pressure PEM electrolyzer. *Int. J. Hydrogen Energy* **2010**. [CrossRef]
57. Zaccara, A.; Petrucciani, A.; Matino, I.; Branca, T.A.; Dettori, S.; Iannino, V.; Colla, V.; Bampaou, M.; Panopoulos, K. Renewable hydrogen production processes for the off-gas valorization in integrated steelworks through hydrogen intensified methane and methanol syntheses. *Metals* **2020**, *10*, 1535. [CrossRef]
58. Yang, S.I.; Choi, D.Y.; Jang, S.C.; Kim, S.H.; Choi, D.K. Hydrogen separation by multi-bed pressure swing adsorption of synthesis gas. *Adsorption* **2008**. [CrossRef]
59. Bernardo, P.; Drioli, E.; Golemme, G. Membrane gas separation: A review/state of the art. *Ind. Eng. Chem. Res.* **2009**. [CrossRef]
60. Bampaou, M.; Panopoulos, K.D.; Papadopoulos, A.I.; Seferlis, P.; Voutetakis, S. An electrochemical hydrogen compression model. *Chem. Eng. Trans.* **2018**, *70*, 1213–1218. [CrossRef]

Article

In Situ Catalytic Methanation of Real Steelworks Gases

Philipp Wolf-Zoellner [1], Ana Roza Medved [1], Markus Lehner [1,*], Nina Kieberger [2] and Katharina Rechberger [3]

[1] Chair of Process Technology and Industrial Environmental Protection, Montanuniversität Leoben, Franz-Josef-Strasse 18, 8700 Leoben, Austria; philipp.wolf-zoellner@unileoben.ac.at (P.W.-Z.); anaroza.medved@gmail.com (A.R.M.)
[2] Voestalpine Stahl GmbH, Research and Development Ironmaking, Voestalpine Straße 3, 4020 Linz, Austria; nina.kieberger@voestalpine.com
[3] K1-MET GmbH, Stahlstraße 14, 4020 Linz, Austria; katharina.rechberger@k1-met.com
* Correspondence: markus.lehner@unileoben.ac.at; Tel.: +43-3842-402-5001

Abstract: The by-product gases from the blast furnace and converter of an integrated steelworks highly contribute to today's global CO_2 emissions. Therefore, the steel industry is working on solutions to utilise these gases as a carbon source for product synthesis in order to reduce the amount of CO_2 that is released into the environment. One possibility is the conversion of CO_2 and CO to synthetic natural gas through methanation. This process is currently extensively researched, as the synthetic natural gas can be directly utilised in the integrated steelworks again, substituting for natural gas. This work addresses the in situ methanation of real steelworks gases in a lab-scaled, three-stage reactor setup, whereby the by-product gases are directly bottled at an integrated steel plant during normal operation, and are not further treated, i.e., by a CO_2 separation step. Therefore, high shares of nitrogen are present in the feed gas for the methanation. Furthermore, due to the catalyst poisons present in the only pre-cleaned steelworks gases, an additional gas-cleaning step based on CuO-coated activated carbon is implemented to prevent an instant catalyst deactivation. Results show that, with the filter included, the steady state methanation of real blast furnace and converter gases can be performed without any noticeable deactivation in the catalyst performance.

Keywords: power-to-gas; catalytic methanation; steelworks; real gases; activated carbon; catalyst poison and degradation

Citation: Wolf-Zoellner, P.; Medved, A.R.; Lehner, M.; Kieberger, N.; Rechberger, K. In Situ Catalytic Methanation of Real Steelworks Gases. *Energies* **2021**, *14*, 8131. https://doi.org/10.3390/en14238131

Academic Editor: Dino Musmarra

Received: 9 November 2021
Accepted: 29 November 2021
Published: 3 December 2021

Publisher's Note: MDPI stays neutral with regard to jurisdictional claims in published maps and institutional affiliations.

Copyright: © 2021 by the authors. Licensee MDPI, Basel, Switzerland. This article is an open access article distributed under the terms and conditions of the Creative Commons Attribution (CC BY) license (https:// creativecommons.org/licenses/by/ 4.0/).

1. Introduction

Integrated steelworks are major contributors to today's global CO_2-emissions. Review publications screening the steelmaking process around the globe revealed that approximately 27 to 30% of any industrial CO_2 emissions are directly linked to this sector [1,2]. With a world-wide crude steel production of 1869 million tonnes in 2019 and a 3.6% per annum average growth rate, the steel demand of our society is increasing strongly. These large amounts of steel are mainly required for building and infrastructure (~52%), mechanical equipment (~16%) and the automotive sector (~12%) [3]. Figure 1 shows the most common route of steelmaking globally, which includes a blast furnace for the reduction of iron ore to hot metal and a converter or basic oxygen furnace for the batch-wise production of molten steel. The accumulating by-product gases, such as the blast furnace gas (BFG), basic oxygen furnace gas (BOFG) and coke oven gas (COG), have a very rich content of CO_2 and carbon monoxide (CO), among other gases (Table 1). At the current stage, these by-product gases are buffered within the steelwork and utilised as an energy carrier internally. Nevertheless, additional fossil energy sources, such as natural gas, are needed to cover the whole energy demand for the power plant and auxiliary energy conversion. Prior to any further use, the product gases are cleaned in a two-stage process, including, for example, a dust collector for the separation of coarse dust, and a venturi scrubber for fine dust and water-soluble components.

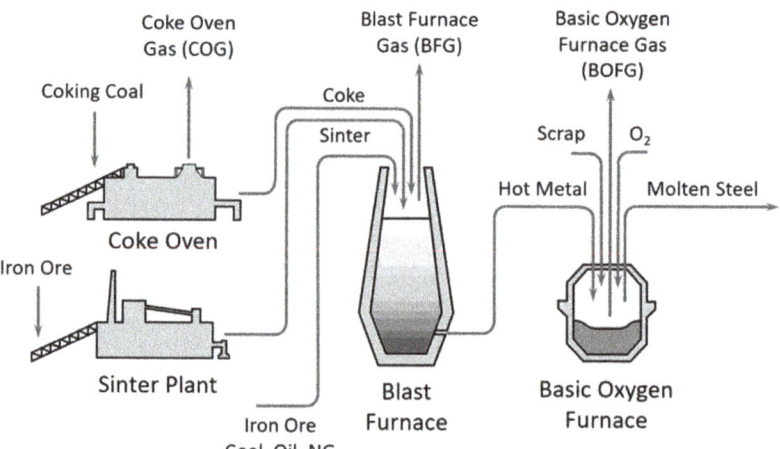

Figure 1. Schematic of steelmaking process—blast furnace/basic oxygen furnace route.

Table 1. Gas composition of by-product gases from a typical steelworks plant [4].

Parameter/Gas Component	Unit	COG		BFG		BOFG
		Min	Max	Min	Max	Mean
CO_2	vol.-% (dry)	1	5.4	16	26	17.2
CO	vol.-% (dry)	3.4	5.8	19	27	60.9
H_2	vol.-% (dry)	36.1	61.7	1	8	4.3
N_2	vol.-% (dry)	1.5	6	44	58	15.5
CH_4	vol.-% (dry)	15.7	27	-	-	0.1
C_nH_m	vol.-% (dry)	1.4	2.4	-	-	-
Lower heating value (LHV)	kJ/Nm^3	9000	19,000	2600	4000	8184

With the challenging targets of the climate agreements being set, the steel industry sector logically seeks for possibilities to reduce their greenhouse gas (GHG) emissions, as well as to incorporate green energy sources in the steelmaking process itself. One way of reducing the GHG emissions, and simultaneously substituting the need for fossil fuels, is the implementation of synthesis processes, such as methanation. In this process, CO_2 and CO react with hydrogen (H_2), gained from green energy sources—for instance, renewable power driving water electrolysis—to create methane (CH_4) and steam (Equations (1) and (2)) [5].

$$CO + 3\,H_2 = CH_4 + H_2O(g) \qquad \Delta H_r^0 = -206\ \text{kJ/mol} \qquad (1)$$

$$CO_2 + 4\,H_2 = CH_4 + 2\,H_2O(g) \qquad \Delta H_r^0 = -165\ \text{kJ/mol} \qquad (2)$$

These two reactions are highly exothermic and are linked via the reverse water–gas shift reaction (Equation (3)).

$$CO_2 + H_2 = CO + H_2O(g) \qquad \Delta H_r^0 = 41\ \text{kJ/mol} \qquad (3)$$

Although these reactions are well-known, the behaviour with real, untreated steelworks gases, as the CO_x source, are yet to be investigated. The detailed fundamentals behind the methanation concept, possible reactor designs and available catalysts are documented by Rönsch et al., combining them with an up-to date overview on methanation projects and the state-of-the-art in the research [6].

Current work on the methanation of steelworks gases primarily focuses on the usage of COG due to its favourable composition. The high amount of hydrogen (up to ~60%)

makes it an attractive feedstock to produce synthetic natural gas (SNG), as it also works as an alternative hydrogen source compared to a solution involving an electrolysing unit (e.g., PEM). Müller et al. [7] investigated the direct conversion of CO and CO_2 from synthetic COG into methane using nickel (Ni)-based catalysts in a fixed-bed reactor. It was shown that additional CO_2 from other sources (e.g., air, flue gas) is required to compensate for the high surplus of hydrogen in the COG to achieve a desirable methane yield. Razzaq et al. [8] tested various Ni-based catalyst support materials (SiO_2, Al_2O_3, ZrO_2 and CeO_2) for the methanation of synthetic COG. The CO_x conversion rates and CH_4 selectivity in a fixed-bed reactor were evaluated. Results showed that ZrO_2-CeO_2-coated catalysts have the highest activity and selectivity for CO and CO_2 for synthetic gases with COG composition.

Medved et al. [9] showed in their work that, although the gas from the coke oven seems to be the favourable by-product gas for methanation, it is already fully energetically integrated into the process chain of integrated steelworks. Due to its high calorific value of up to 19.000 kJ/Nm^3, it is used plant-internally at the blast furnace for firing processes, as well as for the power plant. Consequently, it is not readily available as an input for a methanation unit without significantly affecting the steelmaking process and disturbing the energy balance of the plant. The utilization of COG necessitates its substitution by external energy sources, such as electric power and natural gas, respectively. Therefore, the other two by-product gases, BFG and BOFG, with their very high amount of CO_2, CO and N_2, have been evaluated for their applicability as a feed gas for a methanation plant. Furthermore, the authors concluded that the enrichment of BFG and BOFG via methanation, without the necessity of nitrogen removal, as a lean product gas shows a utilisation potential in the integrated steel plant as a substitute for natural gas. Schöß et al. [10] concluded that, although both gases (BFG and BOFG) are suitable carbon sources for the SNG process with the addition of hydrogen for reaction stoichiometry, the necessary specifications of the natural gas grid cannot be met, due to the high content of nitrogen and the resulting low calorific value. In addition, the significant amount of catalyst poisons present in the already cleaned by-product gases needs to be addressed. Lehner et al. [11] added that, for converting steelworks gases to methane, a load flexible reactor setup is favourable to meet the fluctuations in the process gas and hydrogen availability.

Studies on methanation are mainly based on synthetic gas mixtures simulating real gas compositions. Nevertheless, Müller et al. [12], for example, evaluated the direct CO_2 methanation of flue gases at a lignite power plant, showing that the same commercial Ni-based catalyst used in this work did not degrade during the time frame of the experiments. The real gases included the following catalyst poisons: 63 ppm SO_2, 36 ppm NO_2. The same authors further investigated the CO_2 methanation of flue gases emitted by conventional power plants [13]. They used synthetic gases simulating the real gas composition, including contaminations of up to 100 ppm NO_2 and 80 ppm SO_2. The experiments showed a decrease in the conversion, yield and selectivity by 17% in 12.5 h, or 1.36% per hour. The authors also concluded that the deactivation caused by SO_2 is low in relation to a possible degradation caused by traces of H_2S. Rachow [14] studied the influence of catalyst deactivation by sulphur compounds and NO_x components. The author confirmed through experiments with bottled flue gases from coal-fired power plants, as well as with real gases from the cement industry, that Ni-based catalysts strongly degrade in the range of hours when exposed to SO_2 and sulphur compounds in general. The degree of deactivation depends on the catalyst, its active surface area, the SO_2 concentration and the total volume flow rate. No deactivation was observed during experiments with NO_x contaminations. Méndez-Mateos et al. [15] studied the CO_2 methanation over modified Ni catalysts, integrating promoters (transition metals, such as Mo, Fe, Co or Cr), which were added to the catalyst formulation in different portions. The target was to improve the catalyst's resistance over sulphur, and H_2S in particular. The authors showed that the catalyst activity between 573 and 773 K at 10 bar increased when transition metals were added, except for Mo. In addition, it was possible to regenerate the Co-modified catalyst with oxygen, recovering to a 13% methane yield compared to the fresh catalyst.

Calbry-Muzyka et al. [16] reported on the technical challenges and recent progress made when using biogas as an input for direct methanation. Due to the varying composition of the biogas feedstock, no standard gas cleaning solution has been developed so far. Nevertheless, thorough H_2S and non-H_2S sulphur removal to the sub-ppm level is necessary in order to prevent catalyst deactivation. Witte et al. [17] demonstrated the stable operation of a catalytic direct methanation with biogas in a fluidised bed reactor for over 1100 h. Only a slow deactivation by organic sulphur compounds was identified, which broke through the gas cleaning unit with concentrations between 0.5 and 3 ppmv. The installed unit contained a two-stage adsorption-based biogas cleaning system for deep desulphurization [18]. Fitzharris et al. [19] concluded in their work that Ni-based catalysts are highly sensitive to poisoning by sulphur due to geometric, and not electronic, effects. Concentrations of H_2S as low as 13 ppb in H_2 reduced the steady state methanation activity by more than two orders of magnitude. Bartholomew et al. [20] showed that, under typical low-pressure reaction conditions for methanation (525 K, 1 atm) in a reaction mixture containing 10 ppm H_2S, Ni-based catalysts lost most of their activity within a period of 2 to 3 days. The rate of deactivation due to H_2S poisoning increased with an increasing H_2/CO ratio, as well as with an increasing reaction temperature.

Although the applicability of steelworks gases as feed for the synthesis of methane was addressed and confirmed by multiple authors [7–11], experiments with real gases, including contaminations poisonous to catalysts, have not yet been performed. Consequently, the main aim of this work is to show the degree of degradation when using real BFG and BOFG (including nitrogen) as an input for a methanation unit in a composition, as given in Table 1, as well as to present a working solution to overcome the problem of catalyst deactivation.

2. Methods and Methodology

2.1. Experimental Setup

The experiments presented throughout this paper were carried out with the lab-scaled methanation test rig shown in Figure 2 on the left [21]. The technology was validated in relevant environments, consequently representing a Technology Readiness Level of 5 (TRL 5). The unit consists of three cylindric reactors, each made from austenitic stainless steel (304H chromium-nickel (1.4948)), with a height of 300 mm and an inner diameter of 80 mm. The reactors are connected in series but can also be operated individually (Figure 2, right). At the bottom, each reactor is filled with 3/8″ stoneware balls over a height of 100 mm in order to ensure an evenly distributed gas stream through the catalyst section located on top of the inert material. The catalysts used are either a commercial Ni-based bulk catalyst (Meth 134®, 3–6 mm spheres with a Ni-loading of 20 wt.-%), or Ni/boehmite wash-coated ceramic honeycombs [22]. For the experimental results shown throughout this work, only the bulk catalyst was used. The remaining volume towards the top of the reactors is again filled with the same inert balls.

Figure 3 shows a basic flow chart of the described reactor setup. A maximum flow rate of 3 m³/h (STP, ~50 NL/min) is possible for the input gas stream. Operating pressures of up to 20 bar$_{(abs)}$ can be maintained and the maximum temperature for the reactors is limited to 700 °C. The methanation plant is fed with H_2 (hydrogen 5.0 purity), N_2 (nitrogen 5.0 purity), CO (carbon monoxide 2.0 purity) and CO_2 (BIOGON® C, E290, 99.7% purity) from gas bottles, allowing for the preparation of synthetic gas mixtures to meet the specifications of the by-product gases of interest. In addition, bottled real gases from an integrated steel plant in Austria can be connected to the input stream (additional information provided in Section 2.2). Through thermal mass flow controllers (Bronkhorst), the individual gases enter the gas-mixing station. Before entering the first reactor (R1), the gas stream is preheated in a heat exchanger (W1) to temperatures above 200 °C. Additionally, to reach and keep the required temperature of the catalyst at 260 °C prior to the methanation synthesis, the reactors are equipped with infrared panels (RS Pro, 4 panels per reactor, 500 Watt each) on the outside. Between the reactors, two further heat exchangers (W2, W3) are installed. The final product gas stream is cooled and guided through a condensate trap to extract the H₂O formed during the synthesis. The product gases are combusted in a flare, which is connected to the aspiration system. Four gas sampling stations (at the input, as well as downstream of each reactor) allow for analysis of gas composition with the use of an infrared photometer URAS 26 for CO, CO_2 and CH_4, as well as a thermal conductivity analyser CALDOS 27 for H_2 (both from ABB GmbH) with a deviation of ≤1% per component. The gas analysers are calibrated once every week.

Figure 2. Lab-scale methanation test rig (**left**); reactors and heat exchangers (**right**).

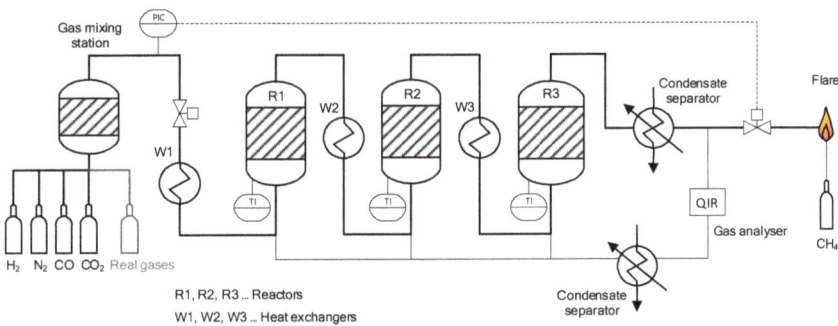

Figure 3. Basic flow sheet of lab-scale methanation plant at Montanuniversität Leoben.

The methanation test rig is equipped with a series of type K thermocouples, as well as pressure and flow rate measurement devices. In addition, a multi-thermoelement is added to each reactor measuring the axial temperature profile 22 mm eccentric from the reactor middle axis. In total, five temperatures are measured inside the catalyst section, as well as two further ones directly below and above the catalyst bed (Figure 4 left) [23]. When using a reactor setup with structured honeycomb catalysts, the locations of the temperature readings are modified to measure the actual temperature inside the channels at the bottom, middle and top of the honeycomb in the centre and at 50% of the reactor's radius (Figure 4 right).

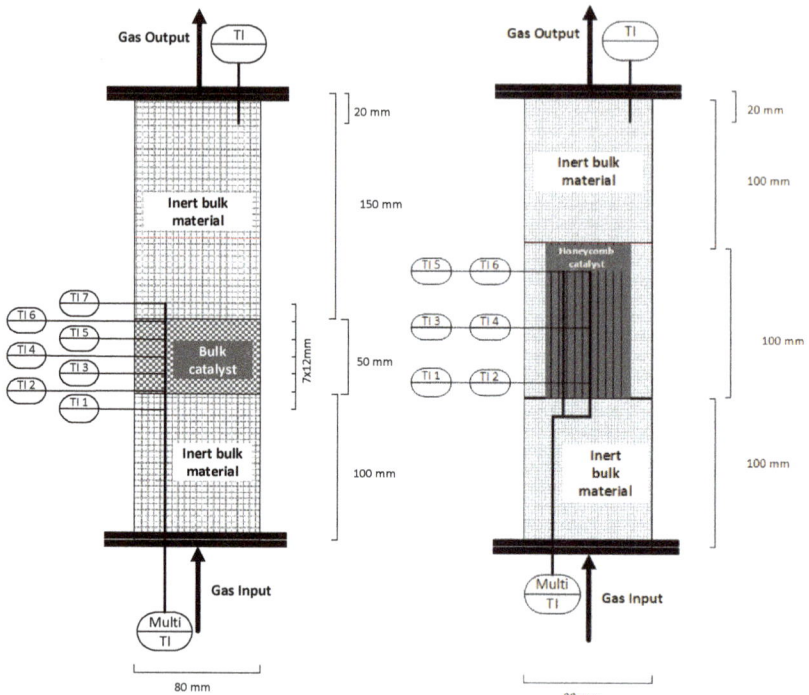

Figure 4. Schematic of one reactor with multi-thermoelement—bulk (**left**), honeycomb (**right**).

The conversion rate of CO and CO_2 as CO_x is calculated based on the feed and product gas composition. The input gas volume flow and the input gas concentrations are known from the mass flow controller setpoints of each species. Whereas CO, CO_2, CH_4 and H_2 can be measured by the gas analysis system at each reactor outlet, the missing species H_2O, as well as the outgoing total gas volume flow, is determined by component and atom balances. The CO_x conversion is then calculated from the ingoing and outcoming component mole flows of CO, CO_2 and CH_4, as shown in Equation (4).

$$CO_x \text{ conversion } [\%] = \frac{n_{in} \times (x_{CO2in} + x_{COin}) - n_{out} \times (x_{CO2out} + x_{COout})}{n_{in} \times (x_{CO2in} + x_{COin})} \quad (4)$$

2.2. Analysis of Bottled Real Gases

Prior to any methanation experiments with real steelworks gases, the content of the provided gas bottles is analysed for gas composition and any catalyst poisons potentially present. These gas bottles are filled within a mobile gas-filling station located in the steel works plant. Compressed BFG or BOFG can be filled into such gas bottles, with a volume of 20 L up to pressures of ~150 bar. For the analysis, a ThermoFisher Trace GC-ultra equipped with three gas channels is used. Hydrocarbons are resolved on a 30 m Rtx-alumina capillary column (ID 0.53 mm; filling Na_2SO_4, 10 µm film thickness) and detected by FID. Permanent gases are resolved on two packed columns, HayeSep Q (2 m × 1/8" OD) and MolSieve 5A (2 m × 1/8" OD) and detected by TCD. Sulphur and phosphorus compounds are determined using an Rtx-Sulphur packed column (2 m × 1/8" OD) and an FPD detector. Helium is used as carrier gas for all three channels.

The analysis of the bottled real gases showed that the samples taken after the gas cleaning station and the gasometers are composed like typical average values in the steel industry [4] (Table 2).

Table 2. Gas composition of bottled real gases from steelworks per gas type.

Gas Component (vol.-% Dry)	BFG	BOFG
N_2	48.6	26.5
O_2	0.6	0.5
CO_2	23.33	17.6
CO	24.17	54.5
CH_4	n.d.	<0.1
H_2	3.3	0.9
$\Sigma\, C_nH_m$	n.d.	<0.1

n.d.—not detectable.

In addition, the following catalyst poisons were detected in the blast furnace gas [24,25]: carbon disulphide (CS_2) with a very small amount of 0.26 mg/Nm^3. COS stabilised at 110 mg/Nm^3, and the SO_2 content was evaluated with 2.2 mg/Nm^3. HCl was below the detectable value of the used equipment (<1.0 ppm), but hydrogen sulphide (H_2S) was detected with values around 28 mg/Nm^3, ammonia (NH_3) with 0.15 and HCN with 0.12 mg/Nm^3. The values for antimony (Sb), mercury (Hg) and other heavy metals that are poisonous to Ni-based catalysts could not be analysed with the selected method. For the converter gas, only small amounts of COS, H_2S and SO_2 were detected; the other catalyst poisons were all below the detectable value of the analysis method. Table 3 summarises the catalyst poisons present in the bottled real gases.

Table 3. Catalyst poisons in bottled real gases for BFG and BOFG [24,25].

Catalyst Poison	BFG	BOFG
	mg/Nm^3	mg/Nm^3
H_2S	28	<1
CS_2	0.26	<1
COS	110	<1
SO_2	2.2	0.99
HCl	<1	0.05
NH_3	0.15	0.05
CH_3SH	<1	<1
C_3H_6O	n.d.	n.d.
HCN	0.12	n.d.
Sb	n.d.	<0.001
Hg	n.d.	n.d.

n.d.—not detectable.

3. Experiments and Results

3.1. Initial Experiments with Bottled Real BFG

For the experiments performed in this work, a reference base case was defined, which serves as a performance comparison between the methanation of synthetic and real gases under steady state and dynamic operating conditions. The experiments have been carried out first with synthetic mixed gas from gas bottles, and additionally with unconditioned and with pre-cleaned real gases from the steel industry.

The parameters of this reference case are:

- H_2-excess rate of 5% to reaction stoichiometry (σ_{H_2} = 1.05, Equation (5));
- Gas hourly space velocity (GHSV, Equation (6)) of 4000 h^{-1} (~16.7 NL/min);
- Operating pressure of 4 bar.

These parameters are based on Medved et al. [9], who analysed the influence of nitrogen on the methanation of synthetic steelworks gases. The authors concluded that a 4–5% H_2 surplus is required within the tested GHSV to achieve a full methane yield for a three-stage methanation setup outlined above. Hauser et al. [26] reported the same value for a heat pipe cooled structured fixed-bed reactor. For the expression of the reaction stoichiometry, the parameter σ_{H_2} is introduced, which describes the ratio of the molar hydrogen flow to the molar flows of CO and CO_2 present in the feed gas (Equation (5)). σ_{H_2} is equal to 1.0 for stoichiometric mixtures, and is <1.0 for under- and >1.0 for over-stoichiometric mixtures, respectively.

$$\sigma_{H_2}[-] = \frac{n_{H_2}}{4 \times n_{CO_2} + 3 \times n_{CO}} \quad (5)$$

The definition of the gas hourly space velocity (GHSV), which is the ratio of the total feed gas volume flow (Q_{gas}) and catalyst volume $V_{catalyst}$, is given in Equation (6).

$$\text{GHSV}\left[h^{-1}\right] = \frac{Q_{gas}}{V_{catalyst}} \quad (6)$$

Figure 5 shows the time-based measurement data for the first experiment with real gases and base case parameters. No gas cleaning system was installed yet. At the starting point, one reactor (R2) was used, and steady-state conditions with a synthetic BFG gas mixture, simulating the real gas composition according to Table 2, were established over a period of two hours prior to the experiment start. The feed gas contained CO, CO_2 and the inert gas N_2, and hydrogen was added to the input stream to reach a surplus of 5% to the reaction stoichiometry (σ_{H_2} = 1.05). At time "0", the bottled real BFG substituted for the synthetic gas mixture, whereby the required hydrogen was kept at σ_{H_2} = 1.05. During the following eight minutes, an immediate and consistent drop in synthesis performance was observed, as the CH_4 content in the product gas stream dropped significantly from ~31 to 24 vol.-%, and, consequently, the CO, CO_2 and H_2 content increased. After a duration of approximately 16 min, the curves started to flatten, but the performance drop remained at a lower rate. This shows that the conversion for the real BFG is less compared to the one with the synthetic gas mixture at the same H_2 surplus, indicating an instant catalyst degradation. As no parameter was varied other than the catalyst poisons, the performance drop can be linked to their presence. This is supported by experimental campaigns performed with synthetic steelworks gases, each with times on stream of 80 h and more (max. 158 h) [9,23,27]. In no case was a catalyst degradation that quick or in this magnitude observed. Although GHSV, σ_{H_2} and the operating pressure were varied for synthetic gas mixtures, the overall performance stayed nearly constant during the whole experiment duration.

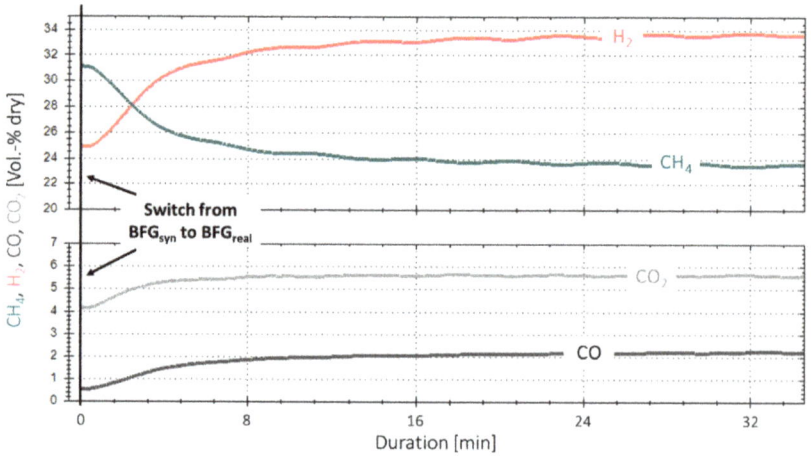

Figure 5. First methanation experiment with real BFG (BFG$_{real}$, 4000 h^{-1}, 4 bar, 5% H_2 excess).

3.2. Variation of Hydrogen Excess Rate at Constant GHSV

With the base case parameters established, experiments with a dynamic H_2 excess rate compared to reaction stoichiometry were performed in order to evaluate the performance of the catalyst and its possible degradation over time. The GHSV, as well as the pressure, were kept constant during these experiments (4000 h^{-1}, 4 bar), and all three available reactors were used. Figure 6 shows the CO_x conversion rates (Equation (4)), as well as the average reactor temperature, for each H_2 excess rate per gas type (real vs. synthetic BFG) and reactor (R1 to R3). The measurements were taken in intervals of 20 min, after which, a steady state of the system, as well as a stable gas sampling, was achieved. Furthermore, the experiment start time for the real gases is plotted at the top. Starting with a σ_{H_2} of 1.05 at 0 min, a slightly lower CO_x conversion rate was measured for R1 after 20 min and for R2 after 40 min, compared to the measurements taken with synthetic BFG (BFG$_{syn}$). Downstream of R3, still full CO_x conversion was achieved with the measurement after 60 min. Afterwards, the H_2 excess rate was increased to 9% (σ_{H_2} = 1.09), which should result in an improved methane yield according to the literature. However, the performance dropped below the one of the 1.05 experiment that started at 0 min. Compared to synthetic gas, no full CO_x conversion could be achieved with all three reactors. The additional experiments with values for σ_{H_2} of 1.02 and 1.0 starting after 120 and 180 min continued the trend towards a significant performance decrease when comparing the real gases with the synthetic gas mixture. This is especially noticeable for the first reactor R1, with a difference of 7.6%-points in CO_x conversion for the σ_{H_2} = 1.0 experiment after approximately three hours. The decrease in catalyst activity is clearly attributable to the catalyst poisons in the methanation feed gases.

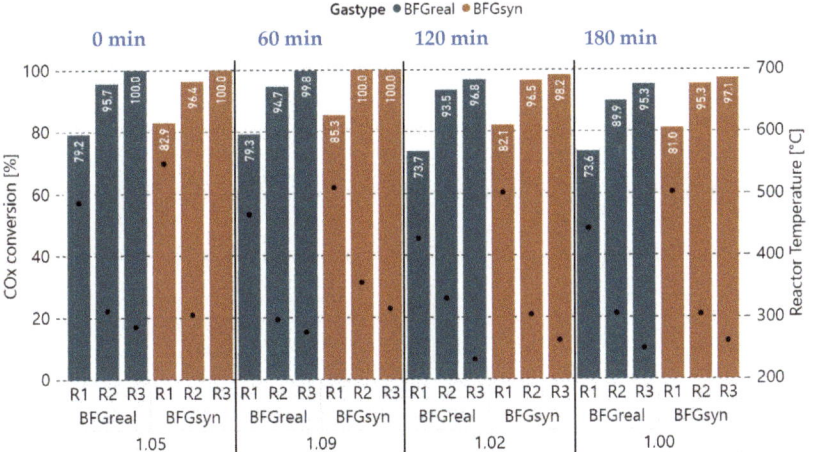

Figure 6. Comparison of experiments with real and synthetic BFG with varying H_2 excess rate (σ_{H_2} = 1.05, 1.09, 1.02, 1.00) at 4000 h^{-1} and 4 bar, CO_x conversion in %, reactor temperature in °C.

Consequently, additional experiments with a 5% H_2 excess rate were performed after this alteration of hydrogen addition to determine the degree of catalyst deactivation. Therefore, the CO_x conversion rates prior to and after 4 h of methanation with real gases (0–240 min) were compared against each other. Figures 7 and 8 show the CO_x conversion and product gas composition per reactor, respectively, for the experiment at 0 min (E-1) and another measurement taken under the same operating conditions after 240 min (E-2). Within a period of 4 h, the performance of the first reactor (R1) dropped by 5.3%-points in CO_x conversion. The second reactor (R2) took over the load as the first reactor's performance dropped, keeping the overall performance of the first two reactors stable. This is confirmed by the temperatures measured inside the reactors, as they decreased for

R1 (median of −71.3 °C) and increased for R2 (median of +35.7 °C). The small drop in CO$_x$ conversion for reactor three (R3) is a result of a too-low average reactor temperature adjusted during experiment E-2.

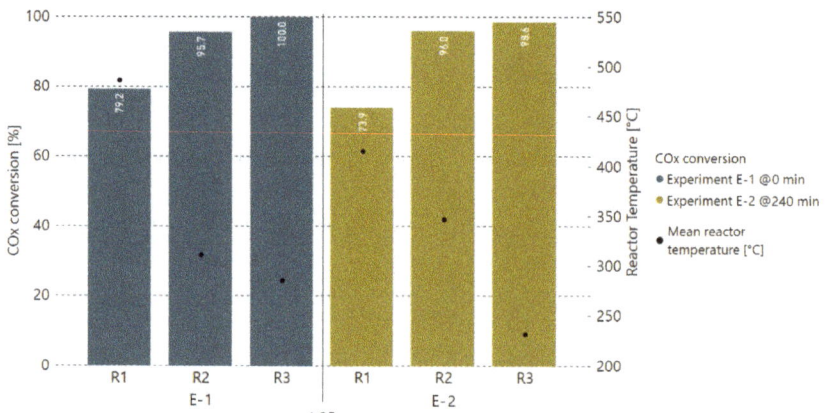

Figure 7. Comparison of CO$_x$ conversion rates in % of two experiments with real BFG, first one (E-1) taken at 0 min, and second one (E-2) taken after 4 h of real gas experiments with varying H$_2$ excess rate.

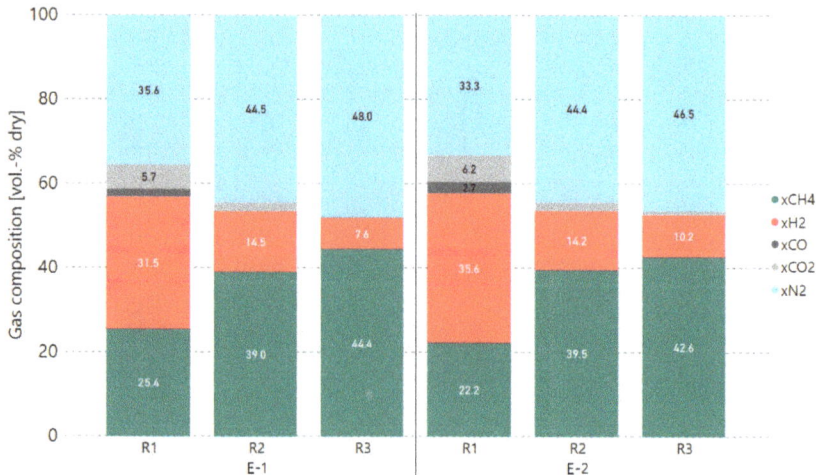

Figure 8. Comparison of real gas experiments E-1 and E-2 with 5% H$_2$ excess rate (4000 h^{-1}, 4 bar), product gas composition in vol.-% dry.

3.3. Extended Experiment Duration with Bottled Real BFG

After the initial tests, a first long-term experiment was performed with bottled real BFG and a hydrogen excess of $\sigma_{H_2} = 1.05$ to further assess the degree of catalyst deactivation. Therefore, base case conditions and parameters were used. It needs to be mentioned that, due to the catalyst deactivation discovered in the previous experiments, only one reactor (R1) was used this time, in order to spare the catalyst in the remaining two reactors (R2 and R3).

Figure 9 shows the results of this extended experiment duration. The product gas composition for the four gases H$_2$, CH$_4$, CO$_2$ and CO is shown in vol.-% dry on the y-axis, and the x-axis shows the duration in hours. Again, a decrease in the overall performance of the catalyst is observed, starting at a higher rate at the beginning that stabilises after 16 min. Over 16.5 h, the CH_4 concentration in the product gas dropped from a starting value of ~29 to ~16 vol.%, indicating the already-mentioned deactivation of the catalyst over time. This is equivalent to a drop in CO_x conversion of ~30% (1.8% per hour).

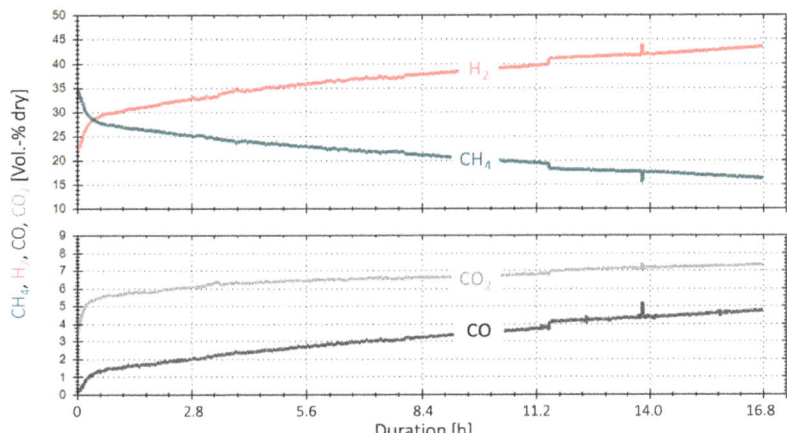

Figure 9. Time-based data for extended experiment duration with real BFG (4000 h^{-1}, 4 bar, 5% H_2-excess) over 16.8 h.

Figure 10 shows the temperature measurements taken inside the catalyst bed of the first reactor (R1) at the beginning of the experiments under synthetic gas conditions (reference base case), and at the end of the last experiment with real gases. The operating conditions are the same (4000 h^{-1}, 4 bar, 5% H_2-excess). A clear shift of the typical bell-shaped temperature curve towards the top of the catalyst bed can be seen. This confirms that the catalyst at the bottom of the first reactor was deactivated by the poisons present in the bottled real gases. This behaviour was not seen for any experiments with synthetic gas mixtures under steady state or dynamic conditions [9,27].

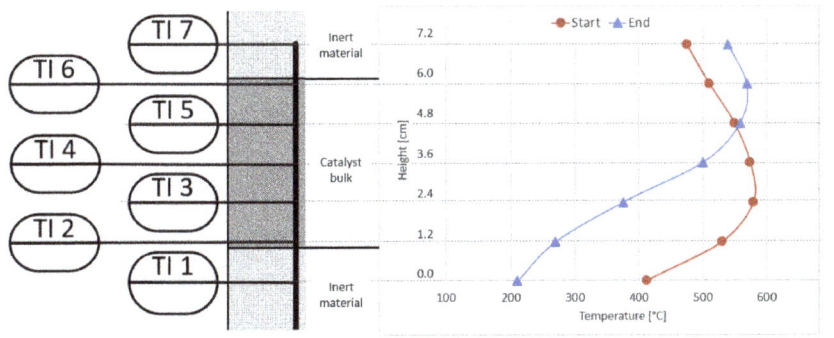

Figure 10. Comparison of reactor R1 temperature profiles at start (red) and end (blue) of long-term experiments with real BFG.

3.4. Analysis of Gas Condensate and Catalyst

Table 4 compares the analysis results of the gas condensate taken after the experiments with real gases with the ones for the synthetic gas mixture. The parameters have been measured according to the norms ISO 10304-1, 17294-2 and 10523.

Table 4. Gas condensate analysis for BFG (real gases vs. synthetic gas mixture).

Parameter	Value Measured for		Unit	Determination Limit
	Real Gas	Synthetic Gas		
pH-value	8.5	6.7	-	-
Chloride Cl	<0.50	<0.50	mg/L	0.50
Sulphate SO_4	<0.50	<0.50	mg/L	0.50
Nickel Ni	73	22	µg/L	1.0
Sulphur S	<5.0	<5.0	mg/L	5.0

The Ni content in the real gases' condensate is more than three times higher compared to the one for the synthetic gas mixture from the reference base case. This confirms the theory that more Ni atoms are taken off the catalyst with real gases and blown out of the reactor setup with the product gas, resp., leaving it through the condensate, indicating a mechanical deactivation through attrition [28]. The decrease in available Ni atoms on the surface of the catalyst can be another reason for the performance drop observed, resulting in a further deactivation of the catalyst. The other parameters relevant for catalyst deactivation (Cl, SO_4, S) were all below the determination limit of the selected analysis methods.

Once the experiments were performed, the possibility to reactivate the catalyst inside the first reactor was evaluated. Therefore, the reactor was purged with pure hydrogen for a duration of 4 h with a GHSV of 2000 h^{-1}, keeping a constant reactor pressure of 4 bar and temperatures above 260 °C. Confirmation experiments afterwards showed no significant improvement of the catalyst's performance for the methanation synthesis. Consequently, the whole methanation test rig was flooded with N_2 to clean the piping and reactors. Afterwards, the catalyst was deactivated with compressed air and withdrawn from the reactors. A sample of the catalyst spheres inside the first reactor was analysed in a scanning electron microscope (SEM) and compared with a new catalyst sample (Figure 11). Just by comparing the two samples visually, a clear change in the surface structure and morphology can be noted. The surface of the used catalyst (right) does not show the crystalline structure of the fresh catalyst anymore (left). SEM reference analyses of used catalyst spheres after 96 h of methanation with synthetic steelworks gases revealed that the surface structure matches the one of the fresh catalyst rather than the one of the real gas experiments, which goes hand in hand with the observation that there was no significant degradation in the overall performance detected during these experiments [9,27]. Due to the high methanation temperatures (up to 600 °C) in the first reactor, thermal sintering is certainly a method of catalyst deactivation that needs to be addressed. Again, the experiments with synthetic steelworks gases and without poisons did not show any signs of a decreased catalyst performance during the experimental campaigns, although maximum reactor temperatures at a similar level were measured. Compared to the real gas experiments, no other parameter than the presence of catalyst poisons was changed. Consequently, this degradation can be clearly attributed to their presence in the real gases.

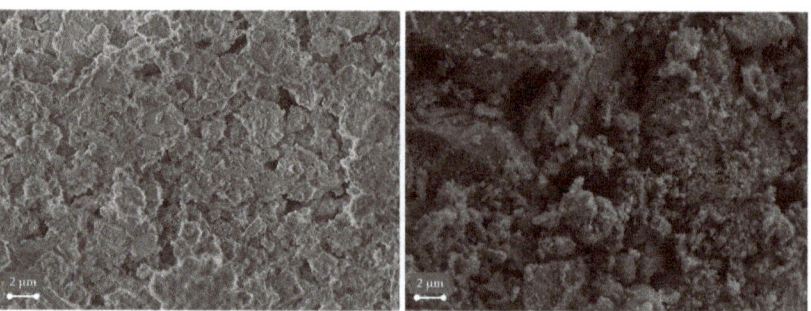

Figure 11. SEM pictures of new (**left**) and used (**right**) bulk catalyst sphere.

3.5. Implementation of Activated Carbon Filter

Based on the results achieved with the direct use of the bottled real gases, including the listed catalyst poisons, which resulted in a quick catalyst degradation, an activated carbon filter was implemented upstream of the first reactor (R1). For this purpose, metal oxide impregnated activated carbon pellets were added to the gas-mixing station. These pellets are specifically developed to remove hydrogen sulphide, organic mercaptans, sulphur dioxide, carbonyl sulphide and nitrogen oxides from oxygen-deficient gas streams, such as CO_2, N_2, CO and H_2. They have a copper content of 7% (as CuO), a diameter of 3 mm, a length of 7 mm, a specific surface area of 936 m^2/g, a pore volume of 0.53 cm^3/g and a bed density of 0.48 g/cm^3 (Figure 12). A total of ~300 mL (187 g) of these pellets was added to the gas mixing station (ID 36 mm, height 300 mm).

Figure 12. Activated carbon pellets with copper oxide coating.

In addition to the implemented filter based on activated carbon, a fresh Ni-based catalyst was added to the reactors. Furthermore, the piping and fittings were exchanged to assure no catalyst poisons remained inside the plant. To test the functionality of the implemented gas cleaning stage with activated carbon, only one reactor was used in the beginning under base case operating parameters. This decision was made to limit the exposure of the plant's components to the real BFG as much as possible, in case of failure of the activated carbon solution.

Figure 13 shows the time-based measurement data for the real gas experiments with one reactor and base case operating conditions. The data include four days of methanation, as well as their individual start-up and shut down phases. Methanation during the first two days (~6 h each) showed very constant measurement values for all gas components on the product side. The actual values vary in a range of ±2%-points around their averages (Table 5) and are within a 1.5%-point range per gas component compared to reference experiments with synthetic BFG, with the same composition and operating conditions. The small fluctuations and peaks, respectively, for H_2 and CO_2, are only process-related due to the measurement technique of the infrared gas analytic station. In addition, the temperature profile is within a ±1 °C range for all temperatures during day one and in a ±2 °C range for all temperatures except TI2 during day two. The temperature at the bottom of the catalyst bulk (TI2) started to decrease after approximately two hours in day two and dropped by 11.1 °C during the remaining four hours of the experiment time. This drop in temperature did not have any influence on the overall catalyst performance, as the CO_x conversion and product gas composition remained stable.

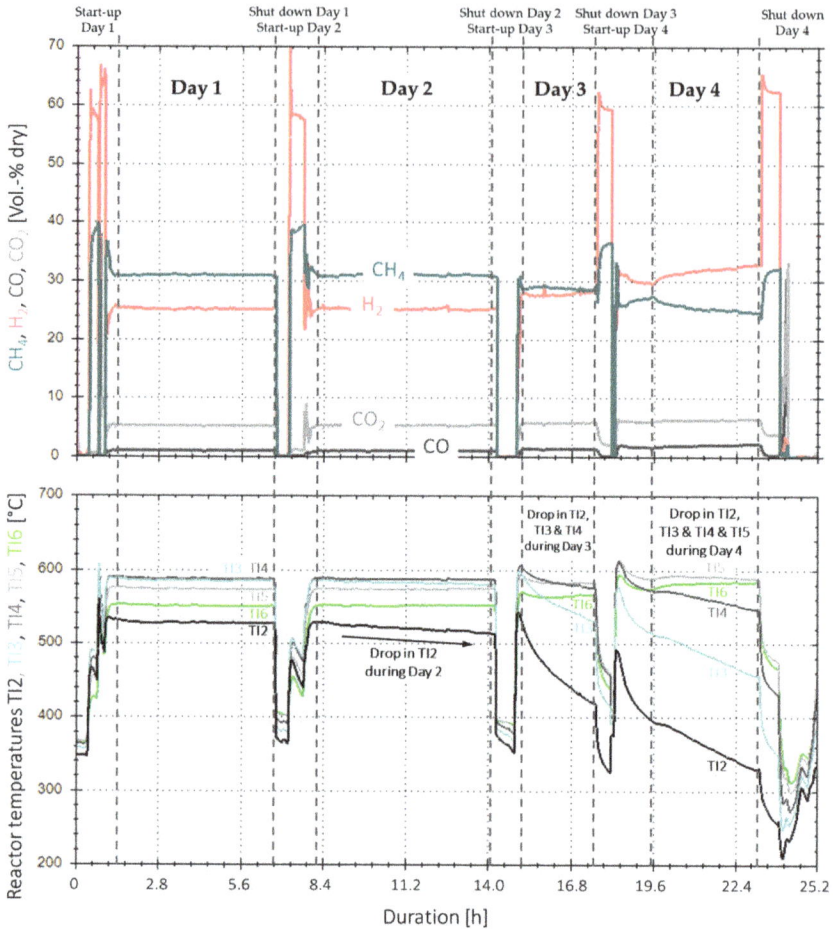

Figure 13. Time-based data for extended experiment duration with real BFG and activated carbon filter (4000 h^{-1}, 4 bar, 5% H$_2$-excess) over 4 days (25.2 h net).

Table 5. Comparison of average product gas composition of methanation with real BFG on day one and synthetic BFG (BFG$_{syn}$); values in vol.-%, activated carbon filter used for real gases.

Gas Component	BFG$_{real}$	BFG$_{syn}$
CO$_2$	5.26	6.0
CO	0.92	0.99
CH$_4$	30.98	32.45
H$_2$	25.19	24.5
N$_2$	37.65	36.06

On day three, the temperature at measurement point TI2 continued its downwards trend with a rate of 38 °C/h. Furthermore, TI3 and TI4 started to decrease by 21.6 °C/h and 8.8 °C/h. respectively, whereas TI5 and TI6 remained stable. Even though the temperatures at the bottom and middle section of the catalyst bulk decreased by 95 and 22 °C in total, the product gas composition remained almost constant. On day four, all temperatures decreased further, except the one for T6, which started to increase as the upper part of the catalyst took over the synthesis load from the catalyst in the lower section. After a net duration of 25.2 h of methanation with real gases, the experiment was stopped, as the temperature at TI2 came close to the lower boundary of 200 °C, below which, poisonous nickel carbonyl is formed [21].

An analysis of the temperature profile along the reactor proves that there is no drop in catalyst performance noticeable on day one, as the red and grey lines in Figure 14 are almost identical. Day two shows the reported minor shift at the bottom of the reactor towards cooler temperatures (yellow line). With day three and four (green and blue lines), the shift towards the left for the lower half of the catalyst bulk is clearly noticeable. At TI5, the temperature remained almost constant during all four days. At the top, the reported temperature increase during the days three and four can be seen, as well as a shift in the hotspot temperature from TI3 to TI5/TI6 over time, clearly indicating the loading of the installed adsorbent bed, as well as a breakthrough of catalyst poisons downstream towards reactor R1.

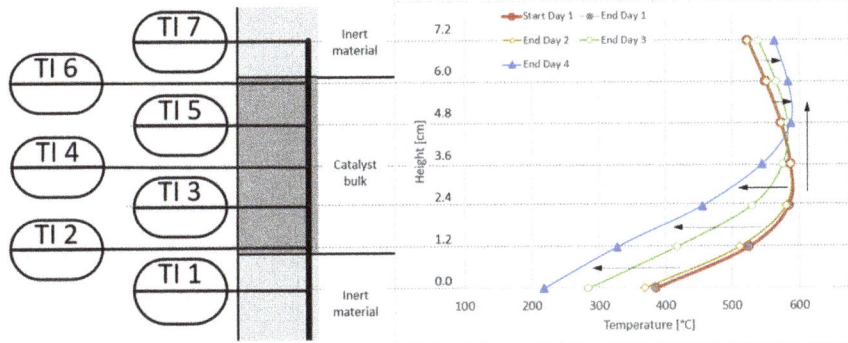

Figure 14. Comparison of reactor R1 temperature profiles at the start of day one, and at the end of each experiment day; activated carbon filter used for real BFG.

Further to the comparison of the temperature profile, reference measurements were also performed prior to and after the real gas experiments on day one and day two. For these, a synthetic gas mixture consisting of hydrogen and CO_2 (H_2:CO_2 = 8:1.5) was used to evaluate if any drop in performance can be recognised. Table 6 lists the product gas composition for these reference measurements. Ref. #1 was taken prior to the real gas experiments, Ref. #2 after day one and Ref. #3 at the end of day two. When exclusively addressing the product gas composition, there is no sign for catalyst deactivation observed until the end of day two.

Table 6. Product gas composition of reference measurements taken with synthetic H_2/CO_2 gas mixture (values in vol.-% dry).

Gas Type	Ref. #1	Ref. #2	Ref. #3
CO_2	0.7	0.9	0.8
CO	0.0	0.0	0.0
CH_4	39.7	39.3	39.5
H_2	57.2	57.7	57.4

Nevertheless, as the temperature profile in Figures 13 and 14 indicates, the first sign of a breakthrough of catalyst poisons occurred on day two after approximately two hours of experiments. This results in a service life of 7.2 h for the implemented adsorbent and is exclusively based on analysing the temperature profile measured inside the reactors. During this time, a total of 7.3 m^3 of real gases flowed through the adsorbent bed, with a volume of 300 mL. Considering the product gas composition, the first catalyst deactivation can be observed after 12.4 h (~12.4 m^3 of real gases). This results in a required amount of absorbent material of 15.1 g per hour of operation with base case parameters, or 15.0 g/m³ of feed gas, respectively.

All of the information obtained from the experiments in lab-scale can be used to estimate and design a pilot plant for the methanation of real gases, including an additional gas cleaning step based on activated carbon. Medved et al. [9] described several application scenarios for the direct methanation of steel gases. One scenario is the substitution of the plant internal natural gas demand with SNG through the methanation of blast furnace gas (BFG). Such a scenario includes approximately 57,000 Nm³/h of BFG and 100,000 Nm³/h of hydrogen. With the consumption numbers for activated carbon obtained in this work, this would result in the need of ~2.4 t of adsorbent per operating hour. Furthermore, Calbry-Muzyka et al. [18] estimated the capacity of the required adsorbent by calculating the integrated loading of H_2S in the adsorbent bed during operation with biogas. With Equation (7), the loading can also be calculated for other catalyst poisons, such as the ones present in the real blast furnace gas. The following parameters are required as an input:

- C_{cat_poison}—avg. molar fraction of catalyst poison in the feed gas [mol/m³];
- Q—avg. flow rate of feed gas stream (real gas plus hydrogen) [m³/h];
- M_{cat_poison}—molar mass of catalyst poison [g/mol];
- t—operating hours [h];
- $m_{adsorbent}$—mass of activated carbon implemented [g].

$$\text{Loading [wt.}-\%] = \frac{\sum C_{cat\_poison} \times Q \times M_{cat\_poison} \times t}{m_{adsorbent}} \quad (7)$$

Table 7 shows the results for a total flow rate of 1.01 m³/h (representing base case parameters), 187 g of adsorbent bed and 12.4 operating hours. This experiment duration was selected as it flags the first sign of catalyst deactivation based on monitoring the product gas composition. Consequently, it also shows a breakthrough of catalyst poisons through the adsorbent bed, which is sufficient for an effect on the catalyst's activity. The loading is below 1.0 wt.-% for each type of catalyst poison measured, with the highest values for COS, followed by H_2S.

Table 7. Loading of adsorbent bed in wt.-% per catalyst poison during methanation with base case parameters over 12.4 h of operation.

Catalyst Poison	BFG$_{real}$	Loading
	mg/Nm³	wt.-%
H_2S	28	0.2
CS_2	0.26	0.01
COS	110	0.73
SO_2	2.2	0.02
HCl	<1	<0.01
NH_3	0.15	0.001
CH_3SH	<1	<0.02
HCN	0.12	0.001

Neubert [29] used a different approach for describing the adsorption of catalyst poisons by comparing the integral change in the axial temperature profiles between two experiments with the one obtained with a fresh catalyst. The author used Equation (8) to calculate the relative activity loss (Δactivity in %) per experiment, which can be further related to the runtime of an experiment (Equation (9)). As a parameter for upscaling, the catalyst consumption ($\Delta m_{catalyst}$) due to sulphur-based catalyst poisons (n_i) can be calculated. This term is multiplied by the amount of fresh catalyst mass ($m_{catalyst,0}$) used in the methanation reactor (Equation (10)).

$$\Delta\text{activity} = \frac{\int_0^z T_m(z) - T_n(z)\,dz}{\int_0^{7.2}(T_1(z) - T_1)\,dz} \quad (8)$$

$$\Delta\text{activity}/h = \frac{\Delta\text{activity}}{\Delta t} \quad (9)$$

$$\Delta m_{catalyst} = \frac{\Delta\text{activity}}{\sum n_i} \times m_{catalyst,0} \quad (10)$$

The results for the four days of experiments with real BFG are shown in Figure 15, as well as in Table 8. During the first two days, the catalyst activity decreased only by less than 3.0%, with a very small catalyst consumption due to sulphur-based catalyst poisons. On day three and four, the catalyst consumption increased to 6.8 and 3.8 $g_{catalyst}/mmol_{sulphur}$, with a combined activity loss of close to 45%. This again shows that, without gas cleaning, respectively, with a fully loaded adsorbent bed, catalyst degradation takes place at an enormous rate. Furthermore, the higher values for Δactivity and $\Delta m_{catalyst}$ on day three compared to day four show that an instant, high drop in performance takes place as soon as poisons break through the adsorbent bed, after which, the decrease stabilises at a lower rate. This behaviour was also observed during the initial tests with real blast furnace gas.

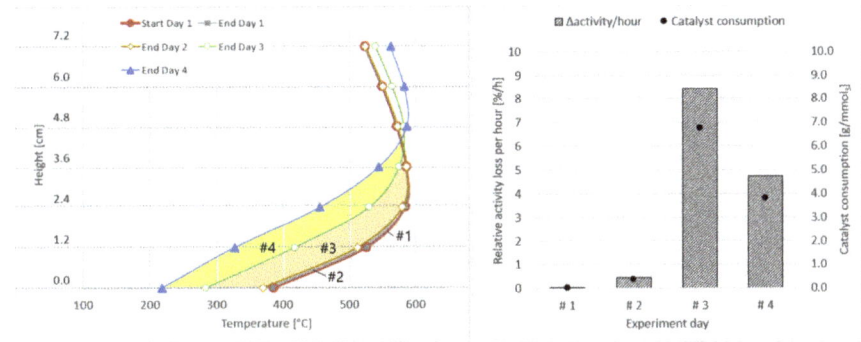

Figure 15. Axial shift of temperature profiles for Δactivity calculation per experiment day (**left**); relative activity loss per hour and catalyst consumption per mmol of sulphur (**right**).

Table 8. Relative activity loss (Δactivity) in % and in %/h, as well as catalyst consumption ($\Delta m_{catalyst}$) in $g_{cat}/mmol_{sulphur}$ and g_{cat}/m^3 for experiments with real BFG (4 days); activated carbon in place.

Experiment Day	Area between Axial Temperature Profiles	Δactivity [%]	Δactivity/h [%/h]	$\Delta m_{catalyst}$ [$g_{cat}/mmol_{sulphur}$]	$\Delta m_{catalyst}$ [g_{cat}/m^3]
#1	Start day 1–End day 1	0.19	0.04	0.03	0.1
#2	End day 1–End day 2	2.79	0.46	0.37	1.0
#3	End day 2–End day 3	21.04	8.42	6.77	18.2
#4	End day 3–End day 4	23.69	4.74	3.81	10.3

When upscaling the figures obtained for day one and day two for the real application scenario in the steelworks, this would result in the deactivation of 88.1 kg of catalyst per hour of operation. Considering the catalyst consumption of the evaluated 12.4 operating hours with a rate of 1.75 g_{cat}/m^3 and upscaling them for the real case, 275 kg_{cat} would be deactivated per hour.

In addition to the experiments with BFG, the upgraded methanation setup, including the activated carbon filter, was also exposed to real gases from a converter (BOFG), with a composition according to Table 2. Again, the required hydrogen to achieve a σ_{H_2} of 1.05 was added through gas bottles. During the experiments with base case parameters and one reactor, no catalyst deactivation was detected over a period of 16 h. As the contaminants of the BOFG are far fewer compared to the ones of BFG (compare Table 3), this is a logical result and shows that the implemented solution works for both gas types. Table 9 compares the product gas composition of methanation with real BOFG and synthetic BOFG. As with blast furnace gas, they are again very similar, varying within 0.5 to 3%-points depending on the gas component.

Table 9. Comparison of average product gas composition of methanation with real and synthetic BOFG (values in vol.-%).

Gas Component	BOFG$_{real}$	BOFG$_{syn}$
CO_2	7.04	6.50
CO	2.28	2.43
CH_4	37.06	38.12
H_2	36.19	38.57
N_2	17.43	14.38

The temperature profile shown in Figure 16, as well as the analysis of the gas condensate, proves that there is no catalyst deactivation noticeable. The temperatures measured inside the reactor at the start and end of the real gas experiments are almost identical and follow the same bell-shaped profile, with a hot spot at measurement point TI4 (middle of the catalyst bulk). In addition, the amount of Ni atoms measured in the gas condensate sample is the same for synthetic and real gases (22 µg/L).

Figure 16. Comparison of reactor R1 temperature profiles at start (red) and end (blue) of experiments; activated carbon filter used for real BOFG.

4. Conclusions and Outlook

In this work, methanation experiments with real gases from the steelworks industry have been performed. These experiments included real by-product gases from the blast furnace, as well as from the basic oxygen furnace (converter), that were directly bottled at an integrated steel plant during normal operation. No further treatment, such as a CO_2 or N_2 separation step, was performed prior to the filling procedure. Methanation without additional gas cleaning resulted in an instant, as well as steady, catalyst degradation due to the poisons present in the real gases. Over the evaluated periods, a drop in the CO_x conversion of ~30%, or 1.8% per hour, was detected for blast furnace gas. The usage of unfiltered real gases resulted in a 3.2 times higher amount of Ni transported out of the reactor setup with the condensate, compared to experiments with synthetic gases meeting the same composition and operating conditions.

As a working solution, an activated carbon filter coated with copper oxide was implemented, showing that a further pre-treatment of the already cleaned steelworks gases is essential prior to feeding them to a catalytic methanation plant. With the activated carbon filter in place, methanation experiments could be performed without any noticeable degradation until the 187 g of activated carbon adsorbent were fully loaded with catalyst poisons. The first signs of deactivation appeared after 7.2 h of operation with real BFG, by means of a drop in the reactor temperature measured at the bottom of the catalyst bulk. Over another period of 5.2 h, the product gas composition and overall conversion rate remained constant, after which, the methane yield started to drop. During this period of 12.4 h, on average, 15.1 g of adsorbent was consumed per hour of operation. The loading of catalyst poisons within the adsorbent bed stayed within a range of 0.01 to 0.73 wt.-% depending on the type of poison. While continuing methanation over another 6.7 h, the catalyst consumption increased from 0.4 to 4.8 $g_{cat}/mmol_{sulphur}$ on average, and the relative activity of the catalyst decreased by ~45% compared to its starting performance. In the case of real BOFG, no signs of catalyst deactivation could be observed during the course of the experiments, which is a result of the far lower catalyst poisons present in this type of gas.

For a real application-based scenario in an integrated steelworks, with the target to substitute the demand of any externally sourced natural gas with a plant-internally produced SNG through methanation, the figures obtained through the experiments at lab-scale would result in the need of ~2.4 t of adsorbent and a deactivation of 88.1 kg of catalyst per hour of operation.

Future work will show the usability of the methanation setup for dynamic experiments as they occur in a steelworks plant, including frequent load changes of up to 25% of gas input power in the range of 5 to 45 min simulating a dynamically operated PEM electrolysing unit. The values obtained through the lab-scaled experiments will also assist in the technical design of a pilot plant for the steelworks industry, including a gas cleaning step based on activated carbon prior to feeding the real gases to the methanation units. Furthermore, the catalyst degradation due to poisons present in the real gases will be investigated in detail through Raman spectroscopy, as well as BET analysis.

Author Contributions: Conceptualization, P.W.-Z. and A.R.M.; Formal analysis, P.W.-Z.; Methodology, P.W.-Z. and A.R.M.; Validation, P.W.-Z., A.R.M., M.L. and N.K.; Visualization, P.W.-Z.; Writing–original draft, P.W.-Z.; Writing–review & editing, P.W.-Z., A.R.M., M.L., N.K. and K.R. All authors have read and agreed to the published version of the manuscript.

Funding: This research was funded by the Research Fund for Coal and Steel RFCS by the EU Commission, grant number 800659 i³upgrade https://www.i3upgrade.eu/ (accessed on 18 August 2021).

Data Availability Statement: Not applicable.

Acknowledgments: The experiments of this work were conducted as part of the research project "i³upgrade—intelligent, integrated, industries", funded by the European Commission. Besides Montanuniversität Leoben, the following research institutes were involved: Chair of Energy Process Engineering (EVT) and Institute of Chemical Reaction Engineering (CRT) at Friedrich-Alexander University Erlangen-Nürnberg, Germany; Central Mining Institute (GIG) in Katowice, Poland; Institute of Communication Information and Perception Technologies (TeCIP) of Scuola Superiore Sant'Anna (SSSA) in Pisa, Italy; and the Centre for Research and Technology Hellas (CERTH), Thessaloniki, Greece; with the industrial partners AIR LIQUIDE Forschung und Entwicklung GmbH (ALFE), voestalpine Stahl GmbH (VAS) and K1-MET GmbH. The authors would also like to acknowledge the work of Hanna Weiss during her bachelor's studies.

Conflicts of Interest: The authors declare no conflict of interest. This paper reflects only the author's view, and the founding sponsors had no role in the design of the study, in the collection, analyses, or interpretation of data, in the writing of the manuscript and in the decision to publish the results. The European Commission is not responsible for any use that may be made of the information contained therein.

Abbreviations

ΔH_r^0	Reaction enthalpy
σ_{H_2}	Ratio of molar hydrogen flow compared to molar flows of CO and CO_2
Q_{gas}	Total feed gas volume flow
$V_{catalyst}$	Catalyst volume
n_i	Molar flows
x_i	Gas composition
C_{cat_poison}	Avg. molar fraction of catalyst poison in the feed gas
M_{cat_poison}	Molar mass of catalyst poison
t	Operating hours
$m_{adsorbent}$	Mass of activated carbon
Δactivity	Relative activity loss
$\Delta m_{catalyst}$	Catalyst consumption
BFG	Blast furnace gas
BFG_{syn}	Synthetic blast furnace gas
BFG_{real}	Real blast furnace gas
BOFG	Basic oxygen furnace gas/converter gas
CH_3SH	Methyl mercaptan
CH_4	Methane
Cl	Chloride
C_nH_m	Higher hydrocarbons (C2+)
CO	Carbon monoxide
CO_2	Carbon dioxide
COG	Coke oven gas
COS	Carbonyl sulphide
CS_2	Carbon disulphide
CuO	Copper oxide
E1,2	Experiment 1 or 2
GHG	Greenhouse gas
GHSV	Gas hourly space velocity
H_2	Hydrogen
H_2O	Water or steam
H_2S	Hydrogen sulphide
HCl	Hydrogen chloride
HCN	Hydrogen cyanide
Hg	Mercury
ID	Inner diameter
ISO	International Organization for Standardization
LHV	Lower heating value
N_2	Nitrogen
NH_3	Ammonia
Ni	Nickel
NO_2	Nitrogen dioxide
NO_x	Nitrogen oxides
OD	Outer diameter
PEM	Proton-exchange membrane
ppm	Parts per million
R1, R2, R3	Reactors 1, 2 or 3
S	Sulphur
Sb	Antimony
SEM	Scanning electron microscope
SNG	Synthetic natural gas
SO_2	Sulphur dioxide
SO_4	Sulphate
STP	Standard temperature and pressure
TI1-7	Temperature indicator 1 to 7

TRL	Technology readiness level
vol.-% (dry)	Share in volume percent (dry basis)
W1, W2, W3	Heat exchanger 1, 2 or 3
wt.-%	Share in weight percent

References

1. Uribe-Soto, W.; Portha, J.-F.; Commenge, J.-M.; Falk, L. A review of thermochemical processes and technologies to use steelworks off-gases. *Renew. Sustain. Energy Rev.* **2017**, *74*, 809–823. [CrossRef]
2. Hasanbeigi, A.; Arens, M.; Cardenas, J.C.R.; Price, L.; Triolo, R. Comparison of carbon dioxide emissions intensity of steel production in China, Germany, Mexico, and the United States. *Resour. Conserv. Recycl.* **2016**, *113*, 127–139. [CrossRef]
3. World Steel Association. World Steel in Figures. 2020. Available online: www.worldsteel.org (accessed on 30 April 2020).
4. Remus, R.; Aguado Monsonet, M.; Roudier, S.; Delgado Sancho, L. Best Available Techniques (BAT) Reference Document for Iron and Steel Production. Joint Research Centre Reference Report. In *Industrial Emissions Directive 2010/75/EU Integrated Pollution Prevention and Control*; European Commission JCR: Luxembourg, 2013.
5. Sabatier, P.; Senderens, J.B. New methane synthesis. *Compt. Rend. Acad. Sci.* **1902**, *134*, 514–516.
6. Rönsch, S.; Schneider, J.; Matthischke, S.; Schlüter, M.; Götz, M.; Lefebvre, J.; Prabhakaran, P.; Bajohr, S. Review on methanation–From fundamentals to current projects. *Fuel* **2016**, *166*, 276–296. [CrossRef]
7. Müller, K.; Rachow, F.; Günther, V.; Schmeisser, D. Methanation of Coke Oven Gas with Nickel-based catalysts. *Int. J. Env. Sci.* **2019**, *4*, 73–79.
8. Razzaq, R.; Zhu, H.; Jiang, L.; Muhammad, U.; Li, C.; Zhang, S. Catalytic Methanation of CO and CO_2 in Coke Oven Gas over Ni–Co/ZrO2–CeO2. *Ind. Eng. Chem. Res.* **2013**, *52*, 2247–2256. [CrossRef]
9. Medved, A.R.; Lehner, M.; Rosenfeld, D.C.; Lindorfer, J.; Rechberger, K. Enrichment of Integrated Steel Plant Process Gases with Implementation of Renewable Energy-Integration of power-to-gas and biomass gasification system in steel production. *Johns. Matthey Technol. Rev.* **2021**, *65*, 453–465. [CrossRef]
10. Schöß, M.A.; Redenius, A.; Turek, T.; Güttel, R. Chemische Speicherung regenerativer elektrischer Energie durch Methanisierung von Prozessgasen aus der Stahlindustrie. *Chem. Ing. Technik.* **2014**, *86*, 734–739. [CrossRef]
11. Lehner, M.; Biegger, P.; Medved, A.R. Power-to-Gas: Die Rolle der chemischen Speicherung in einem Energiesystem mit hohen Anteilen an erneuerbarer Energie. *Elektrotechnik Und Inf.* **2017**, *134*, 246–251. [CrossRef]
12. Müller, K.; Rachow, F.; Israel, J.; Charlafti, E.; Schwiertz, C.; Scmeisser, D. Direct Methanation of Flue Gas at a Lignite Power Plant. *Int. J. Env. Sci.* **2017**, *2*, 425–437.
13. Müller, K.; Fleige, M.; Rachow, F.; Schmeißer, D. Sabatier based CO_2-methanation of Flue Gas Emitted by Conventional Power Plants. *Energy Procedia* **2013**, *40*, 240–248. [CrossRef]
14. Rachow, F. Prozessoptimierung für Die Methanisierung von CO_2-Vom Labor Zum Technikum. Ph.D. Thesis, Brandenburgische TU Cottbus-Senftenberg, Cottbus, Germany, 2017.
15. Méndez-Mateos, D.; Barrio, V.L.; Requies, J.M.; Cambra, J.F. A study of deactivation by H_2S and regeneration of a Ni catalyst supported on Al_2O_3, during methanation of CO_2. Effect of the promoters Co, Cr, Fe and Mo. *RSC Adv.* **2020**, *10*, 16551–16564. [CrossRef]
16. Calbry-Muzyka, A.S.; Schildhauer, T.J. Direct Methanation of Biogas—Technical Challenges and Recent Progress. *Front. Energy Res.* **2020**, *8*. [CrossRef]
17. Witte, J.; Calbry-Muzyka, A.; Wieseler, T.; Hottinger, P.; Biollaz, S.M.; Schildhauer, T.J. Demonstrating direct methanation of real biogas in a fluidised bed reactor. *Appl. Energy* **2019**, *240*, 359–371. [CrossRef]
18. Calbry-Muzyka, A.S.; Gantenbein, A.; Schneebeli, J.; Frei, A.; Knorpp, A.J.; Schildhauer, T.J.; Biollaz, S.M. Deep removal of sulfur and trace organic compounds from biogas to protect a catalytic methanation reactor. *Chem. Eng. J.* **2018**, *360*, 577–590. [CrossRef]
19. Fitzharris, W.D. Sulfur deactivation of nickel methanation catalysts. *J. Catal.* **1982**, *76*, 369–384. [CrossRef]
20. Bartholomew, C.H. Sulfur poisoning of nickel methanation catalysts: I. in situ deactivation by H2S of nickel and nickel bimetallics. *J. Catal.* **1979**, *60*, 257–269. [CrossRef]
21. Biegger, P. Keramische Wabenkatalysatoren zur Chemischen Methanisierung von CO_2. Ph.D. Thesis, Montanuniversität Leoben, Leoben, Austria, 2017.
22. Biegger, P.; Kirchbacher, F.; Medved, A.R.; Miltner, M.; Lehner, M.; Harasek, M. Development of Honeycomb Methanation Catalyst and Its Application in Power to Gas Systems. *Energies* **2018**, *11*, 1679. [CrossRef]
23. Medved, A.R. The Influence of Nitrogen on Catalytic Methanation. Ph.D. Thesis, Montanuniversität Leoben, Leoben, Austria, 2020.
24. Lanzerstorfer, C.; Preitschopf, W.; Neuhold, R.; Feilmayr, C. Emissions and Removal of Gaseous Pollutants from the Top-gas of a Blast Furnace. *ISIJ Int.* **2019**, *59*, 590–595. [CrossRef]
25. Voestalpine Stahl GmbH Prüftechnik & Analytik. Analysenergebnisse der Umwelt-Und Betriebsanalytik. Linz, Austria. 2019. Available online: https://www.voestalpine.com/technischerservice/Prueftechnik-und-Analytik/Umweltanalytik (accessed on 18 August 2021).
26. Hauser, A.; Weitzer, M.; Gunsch, S.; Neubert, M.; Karl, J. Dynamic hydrogen-intensified methanation of synthetic by-product gases from steelworks. *Fuel Process. Technol.* **2021**, *217*, 106701. [CrossRef]

27. Wolf-Zoellner, P.; Lehner, M.; Hauser, A.; Neubert, M.; Karl, J. *i³upgrade–Deliverable D2.4-Demonstration of New Reactors with Real Gases from Steelworks*; Deliverable Report to EU Commission; EU Commission: Brussels, Belgium, 2020.
28. Argyle, M.D.; Bartholomew, C.H. Heterogeneous Catalyst Deactivation and Regeneration: A Review. *Catalysts* **2015**, *5*, 145–269. [CrossRef]
29. Neubert, M. Catalytic Methanation for Small- and Mid-Scale SNG Production. Ph.D. Thesis, Friedrich-Alexander-Universität Erlangen-Nuremberg, Erlangen, Germany, 2019.

Article

Catalytic Hydroisomerisation of Fischer–Tropsch Waxes to Lubricating Oil and Investigation of the Correlation between Its Physical Properties and the Chemical Composition of the Corresponding Fuel Fractions

Philipp Neuner *, David Graf, Heiko Mild and Reinhard Rauch

Karlsruhe Institute of Technology, 76131 Karlsruhe, Germany; david.graf@kit.edu (D.G.); mild.heiko@gmail.com (H.M.); reinhard.rauch@kit.edu (R.R.)
* Correspondence: philipp.neuner@kit.edu

Abstract: Due to environmental concerns, the role of renewable sources for petroleum-based products has become an invaluable research topic. One possibility of achieving this goal is the Fischer–Tropsch synthesis (FTS) based on sustainable raw materials. Those materials include, but are not limited to, synthesis gas from biomass gasification or hydrogen through electrolysis powered by renewable electricity. In recent years, the utilisation of CO_2 as carbon source for FTS was one main R&D topic. This is one of the reasons for its increase in value and the removal of its label as being just exhaust gas. With the heavy product fraction of FTS, referred to as Fischer–Tropsch waxes (FTW), being rather limited in their application, catalytic upgrading can help to increase the economic viability of such a process by converting the waxes to high value transportation fuels and lubricating oils. In this paper, the dewaxing of FTW via hydroisomerisation and hydrocracking was investigated. A three phase fixed bed reactor was used in combination with a zeolitic catalyst with an AEL (SAPO-11) structure and 0.3 wt% platinum (Pt). The desired products were high quality white oils with low cloud points. These products were successfully produced in a one-step catalytic dewaxing process. Within this work, a direct correlation between the physical properties of the white oils and the chemical composition of the simultaneously produced fuel fractions could be established.

Keywords: catalytic dewaxing; hydroprocessing; lubricant production; Fischer–Tropsch

1. Introduction

Environmental trepidations are an excessively discussed topic and the demand for fossil free energy supply is increasing. Current legislative measures, such as the taxation of carbon dioxide (CO_2) as proposed in Germany will inevitably push prices of fossil fuels and carbon sources to novel highs [1,2]. The most common suggestions are the increased usage of electric power in sectors such as transportation and commutation. While an energy supply via battery is viable for cars and light commercial vehicles, it reaches its limits with air and sea travel due to the low energy density of batteries [3]. This problem has reignited the demand for renewable liquid energy carriers, such as middle distillates from Fischer–Tropsch (FT) synthesis. Baseline for this approach is the usage of carbon monoxide (CO) and hydrogen (H_2) as synthesis gas. The original feedstock for FT synthesis, which is coal, can be replaced by renewable sources such as bio mass [4]. Another possibility is the usage or reuse of excess CO_2 as carbon source, which has been proven a viable and flexible alternative for CO [5]. Research has also been conducted regarding the use of the FT synthesis in a power-to-liquid (PTL) approach [6]. A general description of the FT synthesis from CO_2 to paraffins can be defined as the following.

$$CO_2 + H_2 \rightarrow CO + H_2O$$

$$n\,CO + (2n+1)\,H_2 \rightarrow C_nH_{2n+2} + n\,H_2O$$

The main issue with this process is that it will, with certain exceptions [7,8], almost always result in a probability distribution of products according to Schulz–Flory, reaching from methane to heavy paraffinic waxes [9]. The product distribution can be regulated using higher or lower reaction temperatures with lighter products (based on molecular weight) at higher temperatures and heavier products at lower ones [10]. Those processes are known as high temperature FT synthesis (HTFT) and low temperature FT synthesis (LTFT) [11]. The second issue is the paraffinic nature of FT products. Them being free of poly aromatic hydrocarbons (PAH) will generally result in lower emissions and soot formation during combustion [12]. However, the high amount of n-paraffins hinders the application of FT products as fuel due to poor cold flow properties or low octane numbers. A similar issue refers to the wax fraction. With its high melting point, its product applications are rather limited. Examples include industrial coatings and usage in street construction. By lowering the melting point, especially regarding the lubricant sector, more applications can be developed. Those problems can be resolved with a hydroprocessing step. The demand for synthetic lubricants increased as well in recent years due to changing quality demands. Miller et al. reported a shift to lower viscosity oils in order to decrease friction in mechanical stressed areas and lower oil volatility to increase longevity and sustainability [13]. Multiple authors have extensively studied hydroprocessing or hydroconversion of FT products to middle distillates [14–19], making the production of liquid transportation fuels from FT waxes a well-established process. The same cannot be stated about the production of lubricants from FT-waxes. The process itself is established and commercially used, for example, at the Shell PEARL GTL plant in Qatar, which utilizes offshore gas to produce FT products and subsequently lubricants such as technical and medicinal grade white oil [20]. The know-how is, with some mentionable exceptions [21–23], rather limited to patent literature [24–27]. Therefore, the aim of this paper is to investigate and describe the lubricant production from FT waxes and the differences of yield, viscosity and cloud point depending on reaction temperature and pressure. A correlation between molecular composition of the simultaneously produced fuel fractions to the corresponding lube properties is also to be established.

2. Theoretical Background

2.1. Dewaxing

The process of dewaxing describes the removal of n-paraffins from a given hydrocarbon mixture with the purpose of lowering the melting point. There a two main, commercially used dewaxing processes. Solvent dewaxing and catalytic dewaxing. Solvent dewaxing is based on removing wax crystals through precipitation. The untreated oil is usually diluted with an organic solvent such as ketones or aromatics or a mixture of both. It is then cooled below the desired pour point under constant stirring. This results in the formation of wax crystals, which can subsequently be removed by mechanical separation, such as cold filtration [28]. The main advantage of this process is its high yield and high viscosity index during dewaxing of petroleum based feedstock, yet it does require high amounts of extra solvent and additional distillation. Catalytic dewaxing is the selective removal of normal paraffins through cracking or isomerisation. While commercial base stock, such as vacuum gas oil, can be dewaxed by cracking of the long chain paraffins and removal of the lighter fraction via distillation, this is not true for FT-waxes. Those consist almost exclusively of straight chain hydrocarbons. Consequently, cracking would consume the entire wax, transforming it into lighter fuels. To dewax this product, hydroisomerisation is necessary. This will convert the normal paraffins into branched ones by reducing their pour point significantly while mostly maintaining their boiling range [29]. Hsu and Robinson also reported that commercial dewaxing through isomerisation has been available since 1993 when Chevron developed its ISODEWAXING® process using SAPO catalysts. Similar techniques for hydroisomerisation are employed at the Shell PEARL GTL plant.

2.2. Hydroprocessing

Hydroprocessing is an umbrella term for multiple reactions of hydrocarbons under the H_2 atmosphere, including hydrotreating, hydrofinishing, hydrocracking and hydroisomerisation. In this paper, hydroprocessing refers to the latter two. Numerous different mechanisms of hydroisomerisation and hydrocracking were described by Bouchy et al. [15]. An example can be seen in Figure 1. The formation of a new isomer is depicted at (a), while the cracking reaction can be observed at (b). Both require the presence of carbenium ions. The reaction rate of isomerisation is generally higher than the one for cracking, with the notable exception at the scisscion of tertiary carbon ions which can exceed the isomerisation rate.

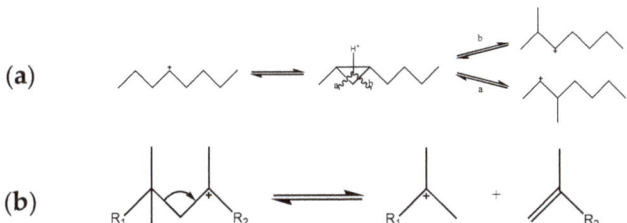

Figure 1. Mechanism of isomerisation (a) and cracking (b) of long-chain paraffins [15].

Cracking of FT products shifts the yield distribution away from waxy components to the desired yields, usually towards the middle distillate range. Isomerisation will lower the melting points in contrast to their n-paraffin counterparts [30], which will result in better cold flow properties.

2.3. Catalytic Mechanism

Hydroprocessing itself is conducted on a bifunctional catalyst, which means there are two different active centres on the catalyst surface, with distinct applications: A metallic site for hydrogenation and dehydrogenation of the product, as well as an acidic site for isomerisation and cracking [15]. The acid component can be provided by a zeolite carrier, which can have varying degrees of acidity. A detailed description of the individual mechanisms is depicted in Figure 2.

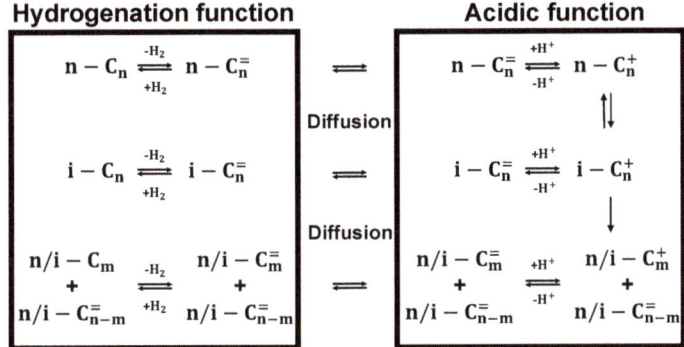

Figure 2. Schematic depiction of the reaction network for hydroprocessing of n-paraffins on a catalytic surface [15].

The saturated n-paraffin is dehydrated on the metallic site and an olefin is formed. The olefin transitions to the acidic site of the catalyst via diffusion. Due to proton uptake on the acidic catalyst site, secondary carbenium ions are formed and those can either crack or branch out to form tertiary carbenium ions [31]. Disintegration of the carbenium ions will result in rehydrogenation and formation of saturated n-paraffins and iso-paraffins.

2.4. Synthetic Lubricants from FT Waxes

Lubricating oils can be described as hydrophobic liquids, which have boiling points higher than water and do not crystalize at room temperature. Commercial Lubricants are specified more clearly. They depend on different parameters including the type of their feedstock (Table 1), their density, viscosity, viscosity index, pour point, cloud point, boiling range, volatility, flash point, amount of saturates, naphthenes and aromatics, oxidative resistance, acidity, colour and conradson carbon residue (CCR). A synthetic lubricant from a GTL process is categorised as API III+ [29].

Table 1. API classifications of lubricant base stock [29].

Group	Manufacturing Process
I	solvent processing
II	hydroprocessing
III	GTL; wax isomerisation, severe hydroprocessing
IV	polyalphaolefins (POA)
V	all other base stocks

Wax from FT synthesis is a decent feedstock for lubricant production due to its low content of heteroatoms or other impurities. The main issue with FTW is that its melting point is usually above 40 °C. The reason is the high amount of linear paraffins in the mixture. In order to reduce the melting temperature of the wax, the chemical composition has to be altered by hydroisomerisation as explained above. FTW dewaxing has been performed using catalytic dewaxing under hydrogen in either a one-step or two-step approach, occasionally including solvent dewaxing as a final step [22–27]. A pretreatment step may necessarily not required. This was stated by Miller et al. [13] who hydroisomerised FT waxes mixed with pyroysed plastics to produce lubricating oils. This results in the conclusion that hydrogenation of pure FT wax prior to hydroisomerisation should be unnecessary. Before hydroprocessing certain wax fractions, what needs to be clarified is whether residual oxygenates can hurt the used catalyst. The two main properties that any lubricating oil will be defined by are its viscosity and its congealing point. The viscosity of hydrocarbon oils is mainly dependent on its average chain length and can only be slightly adjusted by its chemical structure. The kinematic viscosity can even be used to calculate the median molecular weight of the lubricant [32]. The congealing temperature is heavily dependent on the molecular structure, with n-paraffins having the highest melting points in a given carbon range [29].

2.5. Cloud Point

The cloud point (CP) of a paraffin mixture defines the temperature when precipitation of wax crystals begins. It is usually at higher temperatures than other cold flow properties [33]. At CP-temperature, the fluidity of the mixture is not yet inhibited, but wax crystals start to form, which can accumulate on cold surfaces and subsequently plug filters [34]. It is slightly above the common cold filter plugging point (CFPP). Due to it being the highest temperature where significant effects on the consistency of the lubricant can be observed, this work has chosen to display the change in cold flow behaviour.

3. Materials and Methods

3.1. Reactor Setup and Materials Used

In order to apply the processes proposed in the previous section, a three-phase fixed bed reactor was used (Figure 3). The melted wax was kept in a five litre storage vessel from where it was pumped at 90 °C into the reactor by using a high pressure and high temperature piston pump (Bischoff HPD Pump Multitherm 200 model 3351).

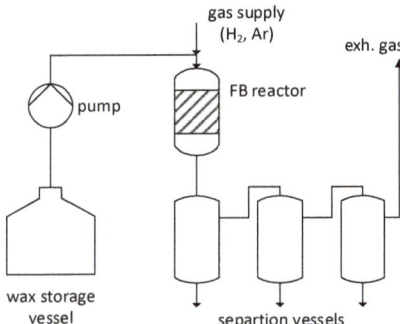

Figure 3. Schematic depiction of the three-phase fixed bed reactor setup at the Engler-Bunte-Institute, Karlsruhe.

This allows a liquid flowrate in the range from 0.6–300 mL/h. The reactor itself consists of a tube with 14.9 mm inner diameter and a length of 800 mm. The used reaction zone itself consists of 400 mm. A 1/8″ tube is inserted in the middle to provide access for temperature measurements. After the reaction zone, the product mixture enters three consecutive vapour–liquid separation vessels with a volume of 800 mL each and from there it can be removed via a needle valve at the bottom. In the reaction zone, 20 g (100–200 µm) of a commercially available zeolitic catalyst with AEL (SAPO-11) structure and 0.3 wt% platinum as a hydrogenation agent were used. This catalyst was chosen with respect to its high selectivity towards isomerisation [35]. The reaction temperature was set on the outside of the reactor tube. Under reaction conditions, the inner temperature increases approximately between 1–2 °C. The feed was Fischer–Tropsch wax from Sasol with a cloud point, as determined by differential scanning calorimetry (DSC) measurement, of 63.6 °C (Figure 4). All experiments were conducted at the same wax flow rate of 0.28 mL/min at 90 °C, resulting in a liquid hourly space velocity (LHSV) of 0.75 h^{-1}. The hydrogen flow of 550 mL/min (at laboratory conditions) was also not changed.

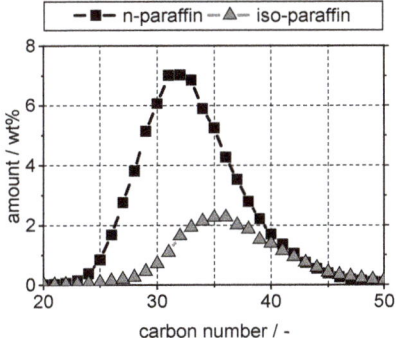

Figure 4. Composition of the feed wax.

3.2. Sample Preparation

The retrieved samples were weighed and separated according to the following atmospheric boiling ranges: gasoline <170 °C, middle distillate 170–340 °C and lubricant >340 °C. In order to avoid overheating the samples, separation of middle distillate from the lubricant was conducted under vacuum using a standard laboratory glass distillation apparatus. The required vacuum was generated with a water-jet pump. The remaining wax crystals were removed by centrifugation (Hettich Rotana 460) at 9000 rpm for 10 min.

3.3. Analytics

The cloud points (CP) of lubricant fractions and middle distillate were determined with differential scanning calorimetry (Netzsch DSC 214 Polyma) under nitrogen atmosphere. In order to achieve exact measurements, a constant cooling rate and sample weight is necessary [36]. For one analysis, 10 mg of the sample was used and a cooling rate of 10 K/min was applied. Afterwards, the sample was reheated before being removed. The cloud point was identified by extrapolating the tangents of the first detectable peak-onset during cooling. The measured temperature at the intersections of both straight lines was noted (T_M). An exemplary peak onset for a commercial paraffin oil and the used temperature programme can be observed in Figure 5. This sequence was developed based on the methods already presented in the literature [37,38] or slightly adapted from similar ASTM Norms [39,40]. The DSC-device required additional calibration to determine the start of crystallisation during cooling. For this purpose, pure components with various solidification temperatures were chosen (n-dodecane, mesitylene, deuterated chloroform, butanone, tetradecane, octacosane, tetracosane, hexadecane and triacontane). Aditionally two Cloud Point reference diesel fuels (ASTM D 2500 [41]) with CPs at 7.7 ± 0.6 °C and −21.0 ± 2 °C were included. The calibration showed significant deviancy of the crystallisation temperature (T_C) and the measured temperature (T_M), the correctional functions can be reviewed in Appendices A–C. The initial calibration did not include the reference fuels. A sample of 10 mg each were analysed. The resulting temperatures were 8.29 °C and −19.09 °C which lies within the stated deviation as determined by ASTM D 2500 and also within the expected scope of DSC measurements, described by Claudy et al. [42]. In order to improve accuracy, the samples were included into the calibration, as presented in Appendix A. It needs to be stated that it was not always possible to determine a clear peak with this method. Some samples crystallized homogenously with no detectable CP. Those were not considered in this work.

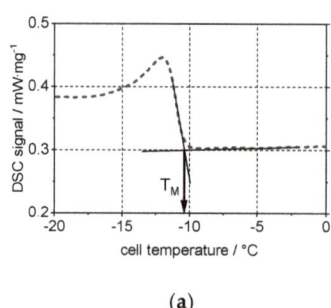
(a)

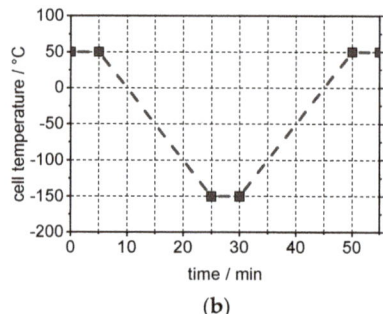
(b)

Figure 5. DSC-Measurement for a commercial paraffin oil (**a**) and temperature programme for determination of cloud points. m_{Sample} = 10 mg, V_{N2} = 100 mL/min (**b**).

The dynamic viscosity was measured with a plate rheometer (Anton Paar MCR 302e) at a sheer rate of 50 s^{-1} and a rotating plate size of 50 mm for five minutes to achieve reliable measurements. The required density to calculate the kinematic viscosity was determined by oscillating tube measurement (Anton Paar DMA 4200). The composition of the gasoline fractions was analysed using a PAC Reformulyzer M4 Hydrocarbon Group Type Analysis. This allowed the measurement of n-/iso-Paraffin content and also detected possible aromatic or naphthenic components. The composition of the produced gas phase was determined by gas chromatography (Agilent Technologies 7890A). Due to calibration problems, the gas phase could only be analysed for one set of parameters (Appendix B) and will not be further discussed in this paper. Determination of the boiling ranges was conducted using a simulated distillation method (Shimadzu Nexis GC-2030).

4. Results
4.1. Boiling Range and Cloud Points

In order to check if the distillation was successful, the boiling range and cloud points of the produced oil had to be measured. This was performed via simulated distillation and an example is given in Figure 6. The individual data points refer to the sum of iso- and n-paraffins within a given carbon number. It was performed in this manner to incorporate response factors. It can be observed that the resulting oil fraction boils entirely above 300 °C and has less than 10 wt% boiling below 340 °C. While the boiling range of the lubricant barely changes, the cloud point could be reduced by 73 °C. With the cloud point being below room temperature and the boiling range indicating no fuel components within the mixture, a lubricating oil was successfully produced. The variation in cloud points was investigated with regards to changing process parameters, in this case the points were reactor temperature and pressure.

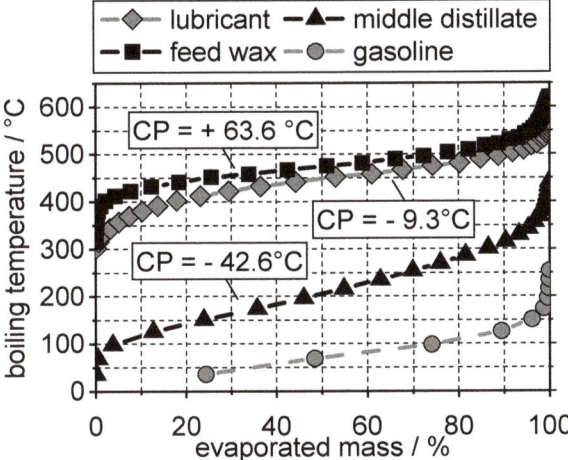

Figure 6. Boiling ranges of the resulting products after hydroprocessing of FT-wax by simulated distillation.

It can be observed that the cloud point decreases rapidly with increasing temperature (Figure 7a). This indicates higher n-paraffin conversion and higher yield of isomers at higher temperatures as expected. Higher pressure results in higher cloud point (Figure 7b). This correlates with the mechanism of hydrocracking, where higher temperature increases carbenium ion formation, while higher pressure inhibits it. The amount of isoparaffins should therefore be higher at low cloud points. A direct measurement of n-paraffins in the oil fraction was not possible with the available setup.

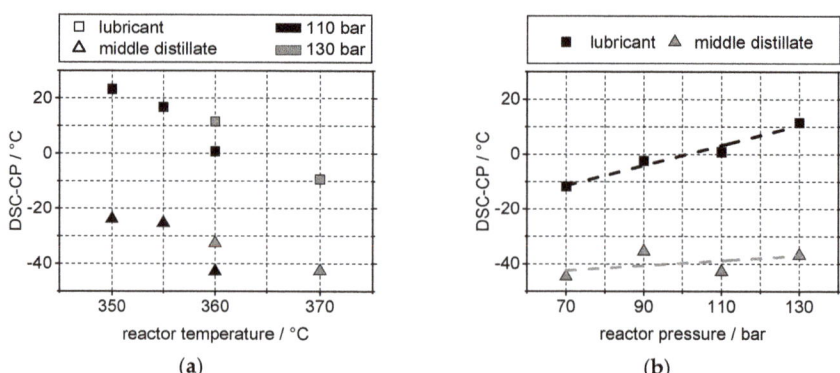

Figure 7. Influence of reactor temperature (**a**) and pressure at T_R = 360 °C (**b**), LHSV = 0.75 h^{-1}, H$_2$/wax = 1964 L/L.

4.2. Yield and Viscosity

Another critical issue is the yield of the potential lubricating oil. The possible liquid yield loss can occur during the reaction itself with the production of high amounts of carbon gasses or during distillation via evaporation to the vacuum pump. The "de facto" yields are calculated in regards to the inlet flow of the FTW and are depicted in Figure 8. The observed trends are similar to the trends witnessed for the cloud points, indicating a dependency on n-paraffin conversion, resulting in the conclusion that low cloud points can only be achieved under loss of yield. Similar effects for the pour point have been reported by Hsu and Robinson [29].

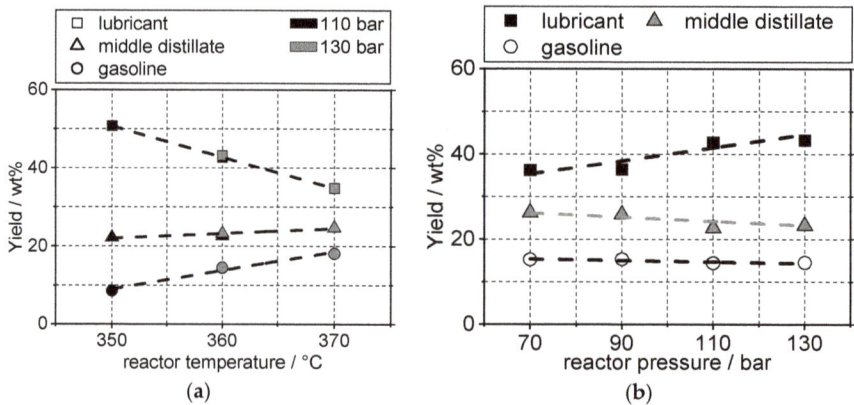

Figure 8. Yield loss at different reactor temperature (**a**) and pressure yield loss at increasing reaction pressure at T_R = 360 °C (**b**); LHSV = 0.75 h^{-1}, H$_2$/wax = 1964 L/L.

To illustrate this issue, the reduction in cloud point was plotted over the potential lubricant yield (Figure 9). Two additional successfully conducted measurements were added (Appendix C, Samples 8 and 9) to verify the following trends at low yield and cloud points. DSC-CP and Yield decrease simultaneously. Higher quality base oil will therefore come at the cost of lower oil return. This effect is independent from reaction temperature or pressure. This concludes a general opposite trend between yield and quality of the base oil. How this dependency will be afflicted by catalyst choice or LHSV variance will be investigated in future studies. It has been shown that catalysts such as ZSM-5 (MFI) generate less lubricant at similar pour points than a SAPO catalyst would [29].

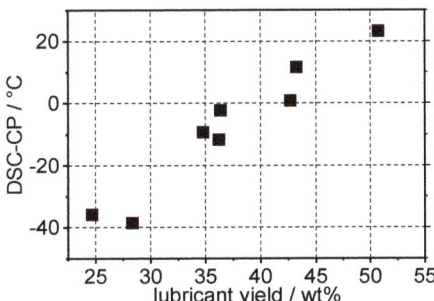

Figure 9. DSC-CP over lubricant yield at varying reactor temperatures and pressure.

While there are clear dependencies between the oil yields and reaction parameters, the same could not be observed for the corresponding viscosity. The fluctuation of viscosity measured at 20 °C was mostly between 41–50 mm^2/s. In the above presented temperature and pressure ranges, barely any trends are observable as depicted in Figure 10. Those trends are not significant enough to have an impact on the oil quality overall. A significant decrease could only be observed for Sample 8 (Appendix C), which might be due to higher catalyst activity for it being produced on a recently regenerated catalyst bed. Yet the measured value for the cloud point of the lubricant and middle distillate was consistent with the expected values, as it will be presented in Section 4.3.

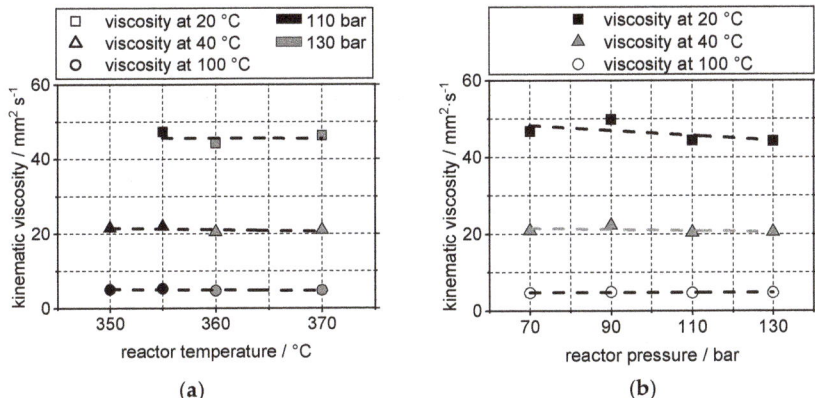

Figure 10. Kinematic viscosity measured at different reactor temperature (**a**) and pressure at T_R = 360 °C (**b**); LHSV = 0.75 h^{-1}, H$_2$/wax = 1964 L/L (viscosity at 20 °C for the point at 350 °C, 110 bar was not determined due to its CP being above that temperature).

4.3. Determination of Cloud Point Using Composition of the Gasoline Fraction

A major issue in lubrication oil analytics is the determination of molecular composition in the mixture. The longer the chain of the molecule, the more isomers and other by-products are possible. A highly isomerised product might not be separable by gas chromatography and requires more sophisticated measurements such as thin film chromatography or high power liquid chromatography. Even if the analytics are available, one major issue is the potential modelling of the fractions with high molecular weight because of unavailable datasets for the produced isomers. On the contrary, the analytics and modelling of products with lower molecular weight has been conducted [43].

The resulting gasoline was analysed regarding its molecular composition and compared to the cloud point of the corresponding lubricant fraction. Figure 11 shows the

DSC-CP for lubricant and middle distillate against the n-paraffin and naphthene content within the conforming gasoline fraction. It can be observed that the cloud point increases simultaneously with n-paraffins in the fuel fraction. The trend is more significant for the lubricants than for the middle distillate. This indicates that the chemical equilibrium of n-paraffins and isoparaffins is comparable within the different boiling ranges. A high n-paraffin content in the gasoline will correspond to a high n-paraffin content in the lube fraction and will therefore begin to crystallise at higher temperatures.

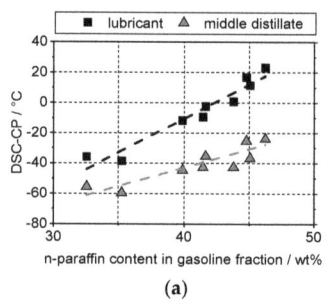 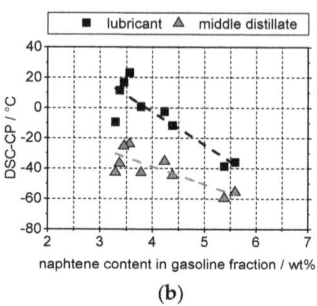

(a) (b)

Figure 11. DSC-CP change of the lubricant and middle distillate fractions against the n-paraffin (**a**) and naphthene (**b**) content in their corresponding gasoline fraction.

A similar but steeper trend can be observed for the naphthenic components. High molecular weight naphthenes are generally decent lubricant components and desirable over n-paraffins for improvement of the cold flow properties.

This method even allows the description of the expected liquid yields (Figure 12). At high n-paraffin content, the highest yields are to be gathered. With the reaction progressing and conversion of n-paraffins through isomerisation and cracking, the yields for gasoline increases, while the middle distillate mostly stays the same.

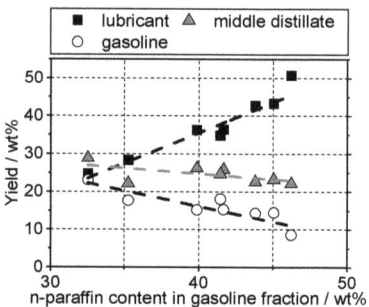

Figure 12. Yield change of the liquid fractions against the n-paraffin content in their corresponding gasoline fraction.

5. Conclusions

In this work, the general production process of base oil from FT waxes was presented. A method for the determination of cloud point using differential scanning calorimetry was applied. It was shown that lubricating oil can be produced by minimal reduction in boiling ranges, while simultaneously reducing the cloud point and subsequently other cold flow properties significantly. It was also shown that the viscosity was hardly reliant on changing reaction parameters. The dependency of cloud points and base-oil yields on reaction parameters, such as temperature and pressure, was investigated and presented. A clear correlation between yield loss and reduction in cloud points could be established. A method for the determination of cloud points through analysis of the distilled gasoline fraction was presented. This would open up different methods for modelling cold flow properties of lubricating oils, while only relying on calculating and analysing the gasoline fraction. These assumptions need to be verified in future studies. In particular, a dependence on different catalyst materials needs to be established.

Author Contributions: The experiments were conceptualised by P.N.; the experimental investigation was performed by P.N., D.G. and H.M.; the manuscript was drafted by P.N.; supervision and funding for this research was provided by R.R. All authors have read and agreed to the published version of the manuscript.

Funding: This research received no external funding.

Data Availability Statement: No application.

Acknowledgments: This paper is dedicated to Hans Schulz on the occasion of his 90th birthday.

Conflicts of Interest: The authors declare no conflict of interest.

Appendix A

Table A1. Calibration equation for CP determination via DSC.

	$T_C = a \cdot T_M + b$ With:	
T_M	a	b
<−12.9 °C	0.7841	0.8216
>−12.9 °C <2.5 °C	0.9900	3.0611
>2.5 °C	0.9701	3.2361

Appendix B

Table A2. Carbon gas composition during hydroconversion of FT-waxes at p_{H2} = 68–75 bar, T_R = 370 °C; LHSV = 0.75 h^{-1}.

Component	Amount
Methane	0.48 wt%
Ethane	0.14 wt%
Propane	5.26 wt%
Butanes	7.10 wt%
n-Butane	03.98 wt%
2-M-Propane	03.13 wt%
Pentanes	3.67 wt%
n-Pentane	01.64 wt%
2-M-Butane	02.02 wt%
Hexanes	1.22 wt%
n-Hexane	00.45 wt%
2-M-Pentane	00.51 wt%
3-M-Pentane	00.26 wt%
Heptanes	0.29 wt%
n-Heptane	00.08 wt%
2-M-Hexane	00.09 wt%
3-M-Hexane	00.12 wt%
Liquid product mix	77.18 wt%
Lubricant (CP = −35.9 °C)	24.68 wt%
Middle distillate (CP = −55.6 °C)	28.88 wt%
Gasoline	23.10 wt%
Distillation loss	00.53 wt%
	95.34 wt%

Appendix C

Table A3. Results table.

Sample	Reactor Temperature (°C)	Reactor Pressure (bar)	Cloud Point (°C)		Reformulyzer Analysis (wt%)			Kinematic Viscosity (mm²/s)				Liquid Yields (wt%)		
			Lubricant	Middle Distillate	Naphthenes	iso-Paraffins	n-Paraffins	20 °C	40 °C	100 °C	Lubricant	Middle Distillate	Gasoline
1	360	70	−11.7	−44.6	4.38	55.60	39.87	46.7	20.9	4.7	36.2	26.2	15.3
2	360	90	−2.4	−35.4	4.23	54.02	41.67	49.8	22.3	4.9	36.4	25.9	15.3
3	360	110	0.8	−42.8	3.78	52.36	43.80	44.4	20.5	4.7	42.7	22.6	14.4
4	350	110	23.3	−23.9	3.56	50.11	46.24	-	21.6	5.0	50.7	22.2	8.6
5	355	110	16.8	−25.3	3.45	49.38	44.77	47.3	22.0	5.3	46.7	22.4	11.5
6	360	130	11.5	−36.8	3.37	51.35	45.06	44.2	20.6	4.7	43.3	23.2	14.5
7	370	130	−9.3	−42.6	3.29	55.20	41.46	46.3	21.1	4.8	34.8	24.7	18.1
8	370	68.75	−35.9	−55.6	5.59	61.72	32.54	33.6	15.7	3.8	24.7	28.9	23.1
9	380	50	−38.6	−59.6	5.38	58.86	35.28	41.1	18.9	4.2	28.4	22.2	17.7

References

1. German Federal Government, Effectively Reducing CO_2 Emissions. 2020. Available online: https://bit.ly/3wivKDO (accessed on 6 July 2021).
2. Eckert, V.; Heinrich, M. German Retail Gas Prices to Rise in 2021 due New CO_2 Tax: Portal. 2020. Available online: https://reut.rs/3yqqAqH (accessed on 6 July 2021).
3. Caspeta, L.; Buijs, N.A.A.; Nielsen, J. The role of biofuels in the future energy supply. *Energy Environ. Sci.* **2013**, *6*, 1077–1082. [CrossRef]
4. Ripfel-Nitsche, K.; Hofbauer, H.; Rauch, R.; Goritschnig, M. BTL—Biomass to liquid (Fischer Tropsch process at the bio-mass gasifier in Güssing). In Proceedings of the 15th European Biomass Conference & Exhibition, Berlin, Germany, 7–11 May 2007.
5. González, M.I.; Eilers, H.; Schaub, G. Flexible Operation of Fixed-Bed Reactors for a Catalytic Fuel Synthesis-CO_2 Hydrogenation as Example Reaction. *Energy Technol.* **2016**, *4*, 90–103. [CrossRef]
6. Pöhlmann, F.; Jess, A. Influence of Syngas Composition on the Kinetics of Fischer-Tropsch Synthesis of using Cobalt as Catalyst. *Energy Technol.* **2015**, *4*, 55–64. [CrossRef]
7. Puskas, I.; Hurlbut, R. Comments about the causes of deviations from the Anderson–Schulz–Flory distribution of the Fischer–Tropsch reaction products. *Catal. Today* **2003**, *84*, 99–109. [CrossRef]
8. Snell, R. Deviations of Fischer-Tropsch products from an Anderson-Schulz-Flory distribution. *Catal. Lett.* **1988**, *1*, 327–330. [CrossRef]
9. Jess, A.; Wasserscheid, P. Chemical Technology. In *An Integral Textbook*; Wiley-VCH: Weinheim, Germany, 2013; ISBN 978-3-527-30446-2.
10. Van der Laan, G.; Beenackers, A.A.C.M. Kinetics and Selectivity of the Fischer–Tropsch Synthesis: A Literature Review. *Catal. Rev.* **1999**, *41*, 255–318. [CrossRef]
11. Leckel, D. Diesel Production from Fischer−Tropsch: The Past, the Present, and New Concepts. *Energy Fuels* **2009**, *23*, 2342–2358. [CrossRef]
12. Gill, S.; Tsolakis, A.; Dearn, K.D.; Fernández, J.R. Combustion characteristics and emissions of Fischer–Tropsch diesel fuels in IC engines. *Prog. Energy Combust. Sci.* **2011**, *37*, 503–523. [CrossRef]
13. Miller, S.J.; Shah, N.; Huffman, G.P. Conversion of Waste Plastic to Lubricating Base Oil. *Energy Fuels* **2005**, *19*, 1580–1586. [CrossRef]
14. De Klerk, A. *Catalysis in the Refining of Fischer-Tropsch Syncrude*; RSC Publishing: Cambridge, UK, 2010; ISBN 978-8-84973-080-8.
15. Bouchy, C.; Hastoy, G.; Guillon, E.; Martens, J.A. Fischer-Tropsch Waxes Upgradingvia Hydrocracking and Selective Hydroisomerization. *Oil Gas Sci. Technol. Rev. l'IFP* **2009**, *64*, 91–112. [CrossRef]
16. Gamba, S.; Pellegrini, L.A.; Calemma, V.; Gambaro, C. Liquid fuels from Fischer–Tropsch wax hydrocracking: Isomer distribution. *Catal. Today* **2010**, *156*, 58–64. [CrossRef]
17. Rossetti, I.; Gambaro, C.; Calemma, V. Hydrocracking of long chain linear paraffins. *Chem. Eng. J.* **2009**, *154*, 295–301. [CrossRef]
18. Schulz, H.F.; Weitkamp, J.H. Zeolite Catalysts. Hydrocracking and Hydroisomerization of n-Dodecane. *Ind. Eng. Chem. Prod. Res. Dev.* **1972**, *11*, 46–53. [CrossRef]
19. Weitkamp, J. Isomerization of long-chain n-alkanes on a Pt/CaY zeolite catalyst. *Ind. Eng. Chem. Prod. Res. Dev.* **1982**, *21*, 550–558. [CrossRef]
20. Royal Dutch Shell plc. Risella X. High-Quality Technical White Oils Based on Gas to Liquids (GTL) Technology. Available online: https://go.shell.com/2RnueBW (accessed on 4 May 2021).
21. Gerasimov, D.N.; Kashin, E.V.; Pigoleva, I.V.; Maslov, I.A.; Fadeev, V.V.; Zaglyadova, S.V. Effect of Zeolite Properties on Dewaxing by Isomerization of Different Hydrocarbon Feedstocks. *Energy Fuels* **2019**, *33*, 3492–3503. [CrossRef]
22. Kobayashi, M.; Saitoh, M.; Togawa, S.; Ishida, K. Branching Structure of Diesel and Lubricant Base Oils Prepared by Isomerization/Hydrocracking of Fischer−Tropsch Waxes and α-Olefins. *Energy Fuels* **2009**, *23*, 513–518. [CrossRef]
23. Kobayashi, M.; Togawa, S.; Ishida, K. Properties and molecular structures of fuel fractions obtained from Hydrocracking / Isomerization of Fischer-Tropsch waxes. *J. Jpn. Pet. Inst.* **2005**, *49*, 194–201. [CrossRef]
24. Kobayashi, M.; Ishida, K.; Saito, M.; Yachi, H. Japan Energy Corporation. Lubricant Base Oil Method of Producing the Same. U.S. Patent 8.012,342 B2, 6 September 2011. Available online: https://bit.ly/3phwTtk (accessed on 3 June 2021).
25. Miller, S.J.; Chevron Research and Technology Company. Wax Isomerization Using Catalyst of Specific Pore Geometry. U.S. Patent 5,246,566, 21 September 1993. Available online: https://bit.ly/3uMNVRf (accessed on 3 June 2021).
26. Bertaux, J.-M.A.; Germaine, G.R.B.; Janssen, M.M.P.; Hoek, A. Shell Internationale Research Maatschappij. Process for producing lubricating base oils. U.S. Patent EP 0776959 A2, 4 June 1997. Available online: https://bit.ly/2S7Az4H (accessed on 3 June 2021).
27. Sirota, E.B.; Johnson, J.W.; Simpson, R.R.; Exxonmobil Research and Engineering Company. Production of Extra Heavy Lube Oils from Fischer-Tropsch Wax. U.S. Patent 7465389B2, 16 December 2008. Available online: https://bit.ly/2SXXNdt (accessed on 3 June 2021).
28. Speight, J.G. *Hydrocarbons from Petroleum*; Elsevier BV: Amsterdam, The Netherlands, 2011; pp. 85–126.
29. Hsu, C.S.; Robinson, P.R. *Petroleum Science and Technology*; Springer Science and Business Media LLC: Berlin/Heidelberg, Germany, 2019.
30. Haynes, W.M. *CRC Handbook of Chemistry and Physics*, 93rd ed.; CRC Press: Boca Raton, FL, USA, 2012; ISBN 978-1-4398-8049-4.
31. Pichler, H.; Schulz, H.; Reitemeyer, H.O.; Weitkamp, J. Über das Hydrokracken gesättigter Kohlenwasserstoffe. *Erdöl und Kohle Erdgas, Petrochemie Vereinigt mit Brennstoff-Chemie* **1972**, *25*, 494–505.

32. ASTM D2502-14. *Standard Test Method for Estimation of Mean Relative Molecular Mass of Petroleum Oils from Viscosity Measurements*; ASTM International: West Conshohocken, PA, USA, 2019. [CrossRef]
33. Coutinho, J.; Mirante, F.; Ribeiro, J.; Sansot, J.; Daridon, J. Cloud and pour points in fuel blends. *Fuel* **2002**, *81*, 963–967. [CrossRef]
34. Dwivedi, G.; Sharma, M. Impact of cold flow properties of biodiesel on engine performance. *Renew. Sustain. Energy Rev.* **2014**, *31*, 650–656. [CrossRef]
35. Deldari, H. Suitable catalysts for hydroisomerization of long-chain normal paraffins. *Appl. Catal. A Gen.* **2005**, *293*, 1–10. [CrossRef]
36. Höhne, G. Problems with the calibration of differential-temperature-scanning-calorimeters. *Thermochim. Acta* **1983**, *69*, 175–197. [CrossRef]
37. Heino, E.-L. Determination of cloud point for petroleum middle distillates by differential scanning calorimetry. *Thermochim. Acta* **1987**, *114*, 125–130. [CrossRef]
38. Rossi, A. Wax and Low Temperature Engine Oil Pumpability. *SAE Tech. Paper Ser.* **1985**, *94*, 768–776. [CrossRef]
39. ASTM D4419-90. *Standard Test Method for Measurement of Transition Temperatures of Petroleum Waxes by Differential Scanning Calorimetry (DSC)*; ASTM International: West Conshohocken, PA, USA, 2015. [CrossRef]
40. ASTM F3418-20. *Standard Test Method for Measurement of Transition Temperatures of Slack Waxes Used in Equine Sports Surfaces by Differential Scanning Calorimetry (DSC)*; ASTM International: West Conshohocken, PA, USA, 2020. [CrossRef]
41. ASTM D2500-17a. *Standard Test Method for Cloud Point of Petroleum Products and Liquid Fuels*; ASTM International: West Conshohocken, PA, USA, 2017. [CrossRef]
42. Claudy, P.; Létoffé, J.-M.; Neff, B.; Damin, B. Diesel fuels: Determination of onset crystallization temperature, pour point and filter plugging point by differential scanning calorimetry. Correlation with standard test methods. *Fuel* **1986**, *65*, 861–864. [CrossRef]
43. Pellegrini, L.; Locatelli, S.; Rasella, S.; Bonomi, S.; Calemma, V. Modeling of Fischer–Tropsch products hydrocracking. *Chem. Eng. Sci.* **2004**, *59*, 4781–4787. [CrossRef]

Article

Combination of b-Fuels and e-Fuels—A Technological Feasibility Study

Katrin Salbrechter [1,*] and Teresa Schubert [2]

[1] Chair of Process Technology and Industrial Environmental Protection, Department of Environmental and Energy Process Engineering, Montanuniversität Leoben, Franz-Josef-Strasse 18, 8700 Leoben, Austria
[2] Research and Development, Wien Energie GmbH, Thomas-Klestil-Platz 14, 1030 Wien, Austria; teresa.schubert@wienenergie.at
* Correspondence: katrin.salbrechter@unileoben.ac.at

Citation: Salbrechter, K.; Schubert, T. Combination of b-Fuels and e-Fuels—A Technological Feasibility Study. *Energies* 2021, 14, 5250. https://doi.org/10.3390/en14175250

Academic Editor: Adam Smoliński

Received: 22 July 2021
Accepted: 19 August 2021
Published: 25 August 2021

Publisher's Note: MDPI stays neutral with regard to jurisdictional claims in published maps and institutional affiliations.

Copyright: © 2021 by the authors. Licensee MDPI, Basel, Switzerland. This article is an open access article distributed under the terms and conditions of the Creative Commons Attribution (CC BY) license (https://creativecommons.org/licenses/by/4.0/).

Abstract: The energy supply in Austria is significantly based on fossil natural gas. Due to the necessary decarbonization of the heat and energy sector, a switch to a green substitute is necessary to limit CO_2 emissions. Especially innovative concepts such as power-to-gas establish the connection between the storage of volatile renewable energy and its conversion into green gases. In this paper, different methanation strategies are applied on syngas from biomass gasification. The investigated syngas compositions range from traditional steam gasification, sorption-enhanced reforming to the innovative CO_2 gasification. As the producer gases show different compositions regarding the H_2/CO_x ratio, three possible methanation strategies (direct, sub-stoichiometric and over-stoichiometric methanation) are defined and assessed with technological evaluation tools for possible future large-scale set-ups consisting of a gasification, an electrolysis and a methanation unit. Due to its relative high share of hydrogen and the high technical maturity of this gasification mode, syngas from steam gasification represents the most promising gas composition for downstream methanation. Sub-stoichiometric operation of this syngas with limited H_2 dosage represents an attractive methanation strategy since the hydrogen utilization is optimized. The overall efficiency of the sub-stoichiometric methanation lies at 59.9%. Determined by laboratory methanation experiments, a share of nearly 17 mol.% of CO_2 needs to be separated to make injection into the natural gas grid possible. A technical feasible alternative, avoiding possible carbon formation in the methanation reactor, is the direct methanation of sorption-enhanced reforming syngas, with an overall process efficiency in large-scale applications of 55.9%.

Keywords: power-to-gas; catalytic methanation; biomass; gasification; synthetic natural gas

1. Introduction

To minimize carbon dioxide (CO_2) emissions and the dependence on energy imports, many European countries see a large potential of biomass gasification for energy or synthetic fuel production. All aspects—heat, power and synthetic fuel production—are regarded in so-called poly-generation concepts, for which gasification represents a key technology [1]. Moreover, biomass is featured with carbon neutrality, which makes clean biomass-based fuel (b-fuel) production through gasification very attractive in future energy systems [2].

Basic considerations of the total process chain of gasification, including up- and downstream process elements, have been discussed in a review by Hofbauer [3], in which the important process principles are explained in detail. A basic flowchart of the process units in gasification processes can be seen in Figure 1. Especially for synthetic fuel or electro-fuel (e-fuel) production, a specific gas composition (e.g., H_2/CO_x ratio) without impurities or catalyst poisons needs to be ensured. However, further extensive gas cleaning and gas upgrading are required in this process route. Biofuels or e-fuels (especially synthetic natural gas (SNG)) can be stored as green energy carriers in existing infrastructure. For the application of syngas in an industrial heat or co-firing process, no specific gas cleaning steps are necessary. In combined heat and power generation, gas cleaning from tar and solid particles is required.

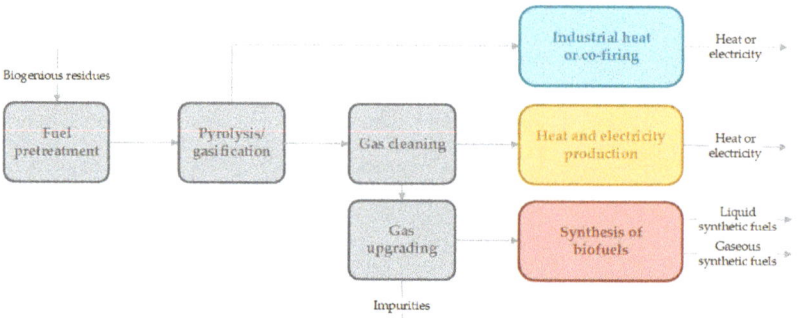

Figure 1. Process units in gasification technology for the production of energy carriers (heat or electricity) or biofuels (Fischer Tropsch products, methanol (liquid) and di-methyl-ether and SNG (gaseous)), inspired by [3], Hofbauer: 2012.

Gasification reactors are distinguished by their fluid mechanical properties in fixed and fluidized bed systems [4]. In fixed-bed gasifiers, mainly coal or waste is used to generate a producer gas. More details on the working principle of fixed-bed gasifiers, as used for example at "Schwarze Pumpe" in Germany, can be found in [5]. In fluidized bed gasification, a bed material—catalytically active or inert—enables very good heat and mass transfer through an equal temperature distribution in the reactor and fuel mixing. The gas used for fluidization also serves as a gasification agent, which can be steam [6], air [7,8] or CO_2 [9–12]. The latter method contributes to a conversion of CO_2 as a carbon capture and utilization (CCU) technology. Dependent on the gas velocity and the particle size distribution of the bed material, two different types of fluidized bed gasifiers exist: bubbling and circulating fluidized beds [13]. The research group of Hofbauer combined these two technologies and developed the dual fluidized bed (DFB) configuration, which offers more advantages for syngas production, such as the production of a nitrogen-free syngas without the need of pure oxygen [3]. Additionally, the gasification and combustion chamber are separated. In the gasification reactor, the bed is fluidized with the gasification agent (steam or CO_2), whereas the combustion chamber is fluidized by air. The generated heat from combustion is transferred to the gasification chamber via a circulating bed material (limestone or olivine) [3]. In this system, both steam and CO_2 gasification processes are being investigated, whereas the latter is a new research topic which helps to reuse CO_2 and enables its conversion to valuable products. Furthermore, sorption-enhanced reforming (SER) in combination with DFB technology is investigated for the production of an adjustable H_2/CO ratio by using limestone as a bed material [14].

Besides feedstock characteristics (which are not considered here), the choice of the gasification set-up, the utilized gasification agent and its operation conditions, such as temperature and pressure, have a strong impact on the produced gas composition. This results in different subsequently needed gas cleaning or gas conditioning systems if the syngas is utilized for methanation processes (see Figure 1). Steam gasification processes generate a nitrogen-free syngas with a high share of hydrogen that matches the requirements for downstream synthesis processes very well. Hence, several demo plants have been realized (GoBiGas in Gothenburg, SWE [15], the Güssing Plant [16], Oberwart [17] and the 1 MW gasifier at the site of Wien Energie [18], all AUT). Syngas from CO_2 gasification, which is still in an early phase of development, shows a high share of CO and CO_2 and only small hydrogen amounts. SER syngas shows a very high share of hydrogen, as a special bed material removes CO_2 from the producer gas through carbonation, but it is not as far developed as steam gasification. The gasification efficiency, known as cold gas efficiency, is the highest in steam gasification, with 84%, and CO_2 gasification and SER feature an efficiency of 73%. This number describes the amount of chemical energy in the product gas in relation to the chemical energy of the fuel introduced in the gasification set-up minus heat losses [19].

Specified data of the gas composition from woody biomass were used from research activities from TU Wien and its 100 kW$_{th}$ dual-fluidized bed gasification pilot plant. Table 1 shows available feed gas compositions with softwood pellets as input material for chemical synthesis of e-fuels [20].

Table 1. Different syngas compositions depending on gasification type based on research activities from TU Wien.

	Column 1	Column 2	Column 3
Source	[20]	[20]	[19]
Species (vol.%$_{db}$)	Steam Gasification	SER	100% CO_2 Gasification
CO	21.2	8.6	40
CO_2	21.5	5.6	40
H_2	48	69.5	15
CH_4	8.8	14	5
C_xH_y	0.5	2.3	0
H_2O	32	41	7
Gasification temperature (°C)	797	629	>840
Bed material	Limestone	Limestone	Olivine
Cold gas efficiency%	84	73	73

The syngas can only be used in biofuel synthesis if potential catalyst poisons such as dust, tars, C_2 species (hydrocarbons with two C atoms), higher hydrocarbons (C_xH_y) and sulfur- and nitrogen-containing species are removed. A detailed summary of different hot and cold gas cleaning methods can be found in Asadullah et al.'s work [21].

A promising route for e-fuel production is catalytic methanation. Biological methanation also describes a well-developed SNG production method, however CO cannot be metabolized by the microorganisms applied in this technology. Therefore, CO shall be separated or converted to CO_2 before the biological methanation. As CO is present in a high share (see Table 1), catalytic methanation is the preferred technology for SNG production from syngas [22]. Catalytic methanation together with electrolysis form the power-to-gas concept, in which electrical energy is converted to chemical energy [23]. The existing natural gas grid offers enormous storage potential for the green gases SNG and H_2 respectively, and enables a link between the usage and seasonal storage of volatile renewable energies. Peak load boilers for district heating supply are run by natural gas, which needs to be substituted to reach the Austrian #mission2030 climate targets [24].

Different reactor set-ups for catalytic methanation processes are described in detail by Kopyscinski et al. [25]. The three basic reactors' set-ups can be classified as follows: fixed-bed reactors (bulk or honeycomb catalysts), fluidized bed reactors or three-phase methanation reactors. In commercial applications, fixed-bed reactors are dominating [26] and can be purchased from different companies [27].

Several reactions [27] play a role in methanation processes. Hydrogenation of CO (Equation (1)) and CO_2 (Equation (3)) aim at the production of methane and water. CO_2 methanation can be seen as a linear combination of CO methanation and a reverse water-gas-shift reaction (rWGS) (Equation (2)). All reaction enthalpies are depicted for 298 K.

$$CO + 3\,H_2 \rightarrow CH_4 + H_2O \qquad \Delta H^R = -206\text{ kJ/mol} \qquad (1)$$

$$CO_2 + H_2 \rightarrow CO + H_2O \qquad \Delta H^R = 41\text{ kJ/mol} \qquad (2)$$

$$CO_2 + 4\,H_2 \rightarrow CH_4 + 2\,H_2O \qquad \Delta H^R = -165\text{ kJ/mol} \qquad (3)$$

As both methanation reactions (Equations (1) and (3)) are volume-reducing reactions, higher pressures favor the production of methane. On the contrary, due to their exothermic nature, lower temperatures improve the conversion rate of CO_x [28]. Detailed illustrative material regarding pressure and temperature dependence of methanation reactions can be found in the publication of Gao et al. [29], where equilibrium compositions are calculated through Gibbs free energy minimization method in CHEMCAD [30].

For syngas methanation, species other than the above-mentioned also need to be considered. Syngas from gasification of carbon species crucially include hydrocarbons, whereas the main representative in the discussed case is C_2H_4. Consequently, many side reactions [29,31] lead to the formation of unwanted by-products, which have a negative effect on methanation performance. Coke formation follows the Boudouard reaction (Equation (4)) and leads to a blockage of active centers on the mainly used Ni-catalysts. Additionally, methane cracking leads to carbon deposition (Equation (6)) at higher temperatures (500–800 °C) [32]. Present hydrocarbons may be hydrogenated to methane (Equation (5)). Produced methane or methane included in the feed gas may undergo Equation (6) and may be cracked. However, formed carbon can undergo steam gasification (Equation (7)) and produce a syngas consisting of a mixture of CO and H_2. Other side reactions are steam- or dry-reforming of ethylene, which both show an endothermic character (Equations (8) and (9)). The reaction enthalpy (at 298 K) of both of the latter mentioned reactions is calculated in HSC 10.

$$2\,CO \rightarrow C + CO_2 \qquad \Delta H^R = -172\text{ kJ/mol} \qquad (4)$$

$$C_2H_4 + 2\,H_2 \rightarrow 2\,CH_4 \qquad \Delta H^R = -202\text{ kJ/mol} \qquad (5)$$

$$CH_4 \rightarrow 2\,H_2 + C_{(s)} \qquad \Delta H^R = 75\text{ kJ/mol} \qquad (6)$$

$$C_{(s)} + 2\,H_2O \rightarrow CO + H_2 \qquad \Delta H^R = 134\text{ kJ/mol} \qquad (7)$$

$$C_2H_4 + 2\,H_2O \rightarrow 2\,CO + 4\,H_2 \qquad \Delta H^R = 289\text{ kJ/mol} \qquad (8)$$

$$C_2H_4 + 2\,CO_2 \rightarrow 4\,CO + 2\,H_2 \qquad \Delta H^R = 292\text{ kJ/mol} \qquad (9)$$

For full conversion, a stoichiometric H_2/CO_x ratio, where $H_2/CO = 3$ for CO methanation and $H_2/CO_2 = 4$ for CO_2 methanation (Equations (1) and (3)), needs to be adjusted in the methanation reactor feed. This combined number is in total defined via the stoichiometric number (SN), which takes both stoichiometric ratios for CO and CO_2 methanation into account (see Equation (10)). At SN = 1, a stoichiometric hydrogen supply is fed into the methanation unit according to Equations (1) and (3).

$$SN = y_{H2}/(3 \times y_{CO} + 4 \times y_{CO2}) \qquad (10)$$

Biomass gasification gained importance around the new millennium due to high subsidies and lower feed-in tariffs for renewable energies. Although biomass shows a lower energy density but a higher inhomogeneity compared to coal, it became an important feedstock for SNG production. Many studies have been carried out regarding the applicability for syngas methanation. Neubert et al. [33] evaluated different methanation possibilities for catalytic methanation from syngas of coal, or biomass gasification including syngas cleaning. Their simulations show that a double-stage process consisting of a structured and a fixed-bed reactor with intermediate condensate separation represents the most reasonable process design option for optimum results. Kienberger et al. [34] dealt with syngas methanation from autothermal fluidized bed gasification, where syngas was used without a pre-cleaning step from tars or sulfur components. They used a common nickel-based catalyst in a polytropic, temperature-controlled reactor. Due to the included impurities, the catalyst consumption increased during methanation processes. Even at the demonstration scale, Rehling [16] reveals that pipeline-ready SNG can be produced in the Güssing 1 MW gasifier, in which softwood was treated with downstream methanation. With improvements of the heat management between the gasification and methanation unit, the overall plant efficiency could be further increased. Basic thermodynamic evaluations of gasification types have been performed by Wang [35]. He proved that steam gasification is the preferred gasification scheme for subsequent biomethane production thanks to the high H_2/CO ratio in the syngas. The performance of the different gasification schemes is evaluated by minimization of Gibbs free energy. Tremel et al. [36] modeled a combination of a small-scale biomass gasification unit with a downstream methanation unit in Aspen Plus. Both process units are realized as fluidized bed systems, and the gasifier needs to be operated at elevated pressures to avoid further compression prior to the methanation reactor. A fully heat-integrated process shows an overall efficiency of 91%, and the SNG quality meets the quality requirements for grid injection. Bartik et al. [31] also examined the combination of biomass gasification with downstream methanation in a fluidized bed, with a focus on low-temperature (300 °C) conversion at ambient pressure, as no compression energy for methanation is needed in this scheme. As gasification product gases, they assessed different syngases from SER, steam and H_2O/CO_2 gasification. A full conversion of CO and CO_2 is only possible for SER product gases, while in the other investigated gas compositions, only full CO methanation could be achieved. Furthermore, H_2O or H_2O/CO_2 gasification produced gases are more vulnerable to carbon formation in the methanation reactor. Steam supply up to 55 vol.% needs to be added for stable operation mode.

In this paper, a specific thermo-chemical production process of SNG in a fixed-bed methanation reactor is investigated. The first step of the evaluated process chain (see Figure 2) is gasification of renewable solid carbon sources with either steam, CO_2 or in the SER process. For catalytic methanation, green hydrogen is assumed to be available, if needed, from water electrolysis powered by renewable electricity to ensure a climate-neutral process and product. The heterogeneously catalyzed methanation process is characterized by Equations (1)–(3). Further fuel upgrading implies the removal of all components to meet requested product quality criteria for feed into the Austrian gas grid, that are specified by the directive ÖVGW G B210 [37].

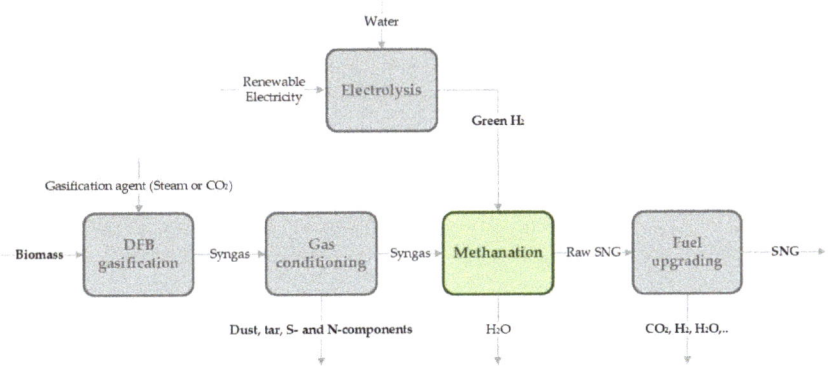

Figure 2. General process route of thermochemical conversion of solid biomass (carbon source) via catalytic methanation to a synthetic fuel (e.g., SNG).

The novelty of the approach presented in this paper is the investigation of different methanation strategies with varying hydrogen supply downstream of the gasification unit, as this has not been considered in the literature so far. Since the syngas obtained from most of the gasification modes lacks a sufficient amount of hydrogen to perform full conversion to SNG, certain amounts of hydrogen have to be added.

The varying hydrogen supply from electrolysis enables a flexible usage of the syngas for e-fuel production depending on hydrogen availability. The limited source of renewable energy must be utilized optimally, and electrolysis capacity represents a costly investment. High costs of around 1100 EUR/kW$_{el}$ for PEM electrolysis are specified in 2020 by Thema et al. [38], following a predicted decreasing trend for the upcoming years. Therefore, this paper aims at investigating possibilities for syngas methanation (methanation strategies) which differ in hydrogen and carbon utilization to determine their technical feasibility, as well as identify promising approaches by evaluating the whole process chain. The detailed targets of this investigation are outlined as follows.

Direct, sub-stoichiometric and over-stoichiometric methanation of different syngas compositions are evaluated in a fixed-bed methanation set-up with thermodynamic considerations, simulation and through experimental tests. Ternary plots will demonstrate the probability of carbon formation during methanation processes and identify necessary remedial actions, such as steam supply, to shift the equilibrium composition. Experimental results applying sub- and over-stoichiometric methanation will be validated with modeled results from simulation works. These strategic considerations provide a future decision-making tool respecting given boundary conditions, such as availability of green electricity for hydrogen production or a limited investment budget. It can be conclusively pointed out which specified process configuration is technically feasible, and which is not.

Direct methanation requires no additionally produced hydrogen from electrolysis, which saves in investment costs and lowers the plant complexity. Due to the lack of hydrogen, a lower conversion rate of included carbon oxides in the syngas is expected and CO_2 separation from the raw SNG is required before a feed into the gas grid. Sub-stoichiometric methanation offers less required electrolysis capacity as hydrogen shall be utilized and converted to a maximum extent. Maximum possible methane output with the provided hydrogen is targeted, taking lower carbon oxide conversion rates into account. Over-stoichiometric methanation enables full conversion of the included carbon oxides thanks to the available excess hydrogen.

Large-scale power-to-gas applications will be surveyed on their technical feasibility as the biomass input power is set to 25 MW, as this represents a large-scale scenario (see [3]). Units such as gasification output, electrolysis capacity and produced SNG from methanation are oriented towards the strategies described above. The effective extent of the biomass to SNG process chain is technically assessed, which allows to evaluate feasibility considerations for possible plant set-ups on a large scale considering available renewable power or investment budget.

2. Materials and Methods

In the following section, possible methanation operating strategies for gasification product gas pursuing different aims are described. Later, the working principle of different technical evaluation tools will be explained. Finally, it will be mentioned which evaluation tool is accordingly assessing each defined methanation strategy and why.

2.1. Different Methanation Operation Strategies (Aim of the Strategy)

- Direct methanation (low-investment strategy)

Generated syngas gas would be directly fed into the methanation unit without additional hydrogen. Direct methanation follows a low-investment strategy as the necessity for the construction of an electrolysis is not provided in this case.

This strategy is applied to all depicted syngas compositions in Table 1 (for product gas from steam gasification, from SER and from 100% CO_2 gasification) to evaluate whether the included share of hydrogen is sufficient for downstream methanation and if carbon formation may occur during fuel synthesis.

- Sub-stoichiometric methanation at SN < 1 (maximum hydrogen usage)

As hydrogen is a valuable input in methanation processes because of its high specific production costs, its consumption is minimized, pursuing the smallest share of hydrogen present in the raw synthetic natural gas after methanation, while maximum carbon conversion is targeted. More specifically, the SN will be lowered in the feed to a selected sub-stoichiometric proportion, which ensures a maximum usage of the available hydrogen from electrolysis, while the highest possible carbon utilization in methanation is also considered at the same time. Through this operation strategy, hydrogen demand, and therefore electrolysis capacity, investment and renewable electricity consumption, can be reduced to a minimum level, which on the other hand ensures optimized carbon conversion.

This strategy is applied to product gas from steam gasification in the experimental evaluation section, as this syngas composition strikes the best balance between included hydrogen in the syngas and additionally needed hydrogen produced via electrolysis. SER offers a very high share (70 vol.%) and CO_2 gasification a too-low (5 vol.%) share of hydrogen in the syngas. Considering the SER syngas composition, hydrogen is already present in a widely over-stoichiometric ratio (SN > 1). In the case of 100% CO_2 gasification, the syngas offers a very low hydrogen share (SN = 0.05), requiring a significant large electrolysis capacity. Due to these extremely different hydrogen shares (either too high or very little), these syngas compositions are not considered in experimental tests pursuing sub-stoichiometric methanation.

- Over-stoichiometric methanation at SN > 1 (maximum carbon usage)

For the desired full conversion of CO_x, sufficient hydrogen needs to be present. Since syngas contains a small share of methane, the chemical equilibrium shifts to the educt side in methanation processes as one of the reaction products is included in the feed. A slight surplus of hydrogen in the feed for methanation will shift the equilibrium back to the product side, enables full conversion of the present CO_x components in the syngas, limits carbon deposition and achieves high selectivity for methane. A technologically evaluated hydrogen excess lies at 3% above methanation stoichiometry, as it has been assessed by Krammer et al. [39] for the production of a satisfying product gas composition.

This strategy will only be applied to steam gasification syngas in the experimental section following the same arguments as mentioned in the section on the sub-stoichiometric methanation strategy.

To summarize the approach presented here, an overview about the applied methanation strategies on different syngas compositions is shown in Table 2.

Table 2. Overview of applied methanation strategies on different syngas types with additional hydrogen supply from electrolysis (*).

		Type of Syngas		
		Steam Gasification	SER	100% CO_2 Gasification
Methanation Strategy	Direct	x	x	x
	Sub-stoichiometric (SN < 1)	x (*)	30% hydrogen excess available in raw product gas from gasification	Very little hydrogen available in raw product gas from gasification
	Over-stoichiometric (SN > 1)	x (*)		

2.2. Technical Evaluation Tools

- Basic thermodynamic evaluations

Fundamental thermodynamic evaluations of Gibbs free energy (ΔG) were conducted with the software HSC 10, as the minimization of ΔG features chemical equilibrium and predicts occurring reactions during methanation processes, assuming ideal gas mixtures. Equations (1)–(9) will be assessed.

- Ternary plot

With the help of a ternary diagram, the investigated syngas composition can be shown as a single point at chemical equilibrium in a 2D coordinate system described by the C-H-O-ratio. Six species which appear in Equations (1)–(4) are constituted of three atoms, namely C, H and O. Equilibrium lines of carbon deposition (depending on selected temperature and pressure) will be implemented, dividing the area in which carbon deposition is thermodynamically possible (above lines) or not (below lines). The ternary plot visualizes if carbon deposition occurs for each specific syngas composition. Frick et al. [40] and Bai et al. [41] also used ternary diagrams for visualizing possible carbon formation in methanation processes and approved that ternary diagrams are an adequate tool for the design-finding procedure of the methanation section.

- Determination of the optimum sub-stoichiometric hydrogen feed

Aspen Plus was used to model the syngas conversion in a catalytic methanation process and to predict the raw SNG composition. To find the optimum hydrogen supply for sub-stoichiometric methanation, two Gibbs reactors assuming thermodynamic equilibrium by minimization of Gibbs free energy were applied. In this simulation, the lowest possible hydrogen feed is modeled chasing two different but coupled goals. Firstly, the least possible share of hydrogen in the product gas should be obtained. Secondly, the available hydrogen is supposed to convert itself with as many carbon oxides as possible. This results in the main goal of this strategy, which is described by a full conversion of the valuable resource, hydrogen, because of its costly and energy-intensive production.

- Lab-scale experiments

As an experimental methanation set-up, the laboratory reactor cascade at the Chair of Process Technology in Leoben was used (see Figure 3). The pilot plant consists of three fixed-bed reactors, which can be operated alone or in series, and each of them is filled with 0.25 L of a commercial 20 wt.% Ni-bulk catalyst named Meth 134®. To examine the methanation performance of syngas from steam gasification, two fixed-bed reactors were operated in series. The first reactor stage reaches thermodynamic equilibrium and hinders the full conversion of CO_x. Gas mixtures, synthetically mixed from gas bottles, according to Table 1, can be fed with up to 50 L_{STP}/min, and a maximum pressure of 20 bar. Gas cooling between the reactor stages is attained through uninsulated pipelines, so that condensate is drained at the lowest pipeline height before the inlet to the next reactor stage. A multi-thermocouple with six measuring points along the axis in the catalyst bed enables accurate temperature measurement in each reactor, which is schematically pictured in Figure 3. The gas composition of the intermediate product (after the first methanation stage) and the final product (after the second methanation stage) was analyzed with an infrared photometer (AL3000 URAS26) and a thermal conductivity analyzer (AL3000 CALDOS27) from ABB [28,42].

In Table 3, the synthesized gas mixtures used for experimental test runs are shown, considering both a hydrogen feed at SN = 0.78 and SN = 1.03 for steam gasification product gas. For test runs, dry gas mixtures are produced by bottled synthesized gases, and hydrocarbons are not available to be fed into the lab-scale methanation plant. The operating pressure was set to 7 bar since this pressure level strikes a good balance between necessary syngas compression and sufficient methanation performance, as CO and CO_2 hydrogenation (Equations (1) and (3)) are strongly pressure-dependent [39]. The gas flow amounts to 8.4 L_{STP}/min, resulting in a GSHV (gas hourly space velocity) of 2000 h^{-1}. The GHSV is calculated as the value of standard volume input flow divided by the catalyst volume.

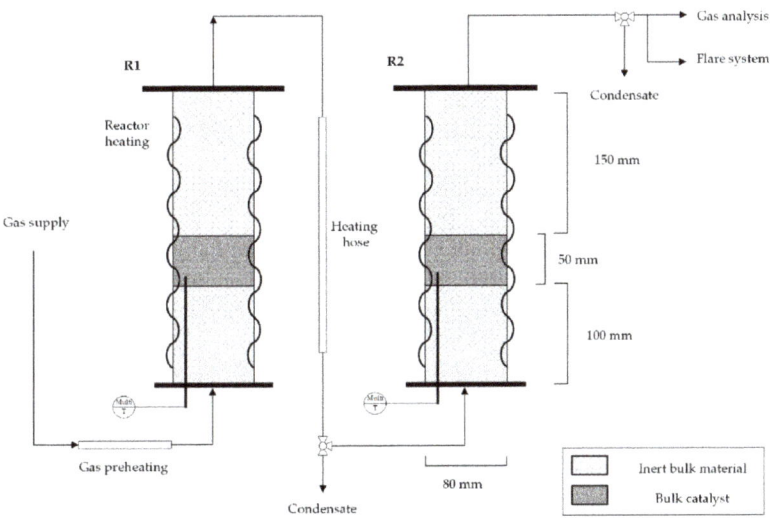

Figure 3. Laboratory double-stage methanation set-up consisting of two fixed-bed reactors (R1 and R2) with an introduced multi-thermocouple in the catalyst bed.

Table 3. Steam gasification syngas composition used for experimental lab-scale methanation test runs at 7 bar and at GHSV = 2000 h^{-1}. Hydrogen included in the syngas and additional supplied hydrogen are indicated.

Gas Type	Syngas from Steam Gasification	
Methanation Strategy	Sub-Stoichiometric	Over-Stoichiometric
SN	0.78	1.03
Dry syngas composition in molar share in %	–	–
CO	12.6	10.3
CO_2	12.9	10.5
H_2	28.1	22.9
CH_4	4.9	4.1
Additional H_2	41.5	52.2
Total %	100	100

- Modeling approach for lab-scale methanation results

Experimental results were also modeled in Aspen Plus by implementing a kinetic model from Rönsch et al. [43], as the model allows a broad temperature range due to its background from dynamic methanation operation. In this model, CO_2 methanation is regarded as the linear combination of CO methanation and rWGS. Rönsch et al. propose two reaction rates for CO methanation for an 18 or 50 wt.% Ni-catalyst. In the present work, the kinetic reaction rate using the 18 wt.% Ni-catalyst (published by Klose et al. [44]) was chosen as it matches best with the catalyst implemented in the laboratory set-up. The implemented reaction rate by Rönsch et al. follows the format of a LHHW (Langmuir–Hinshelwood–Hougon–Watson) approach. The reactor set-up in the simulation flowsheet is modeled as a one-dimensional plug-flow reactor system with two stages (Figure 4). This layout best represents the experimental set-up consisting of two fixed-bed reactors in series, considering the measured temperature profiles in each reactor. For the modeling process, experimental parameters such as the reactor dimensions (d_i = 80 mm), chosen catalyst (20 wt.% Ni-loading) with a height of 50 mm, temperature profiles, pressure level (7 bar) and the investigated feed gas composition from Table 3 were implemented in the chosen kinetic model.

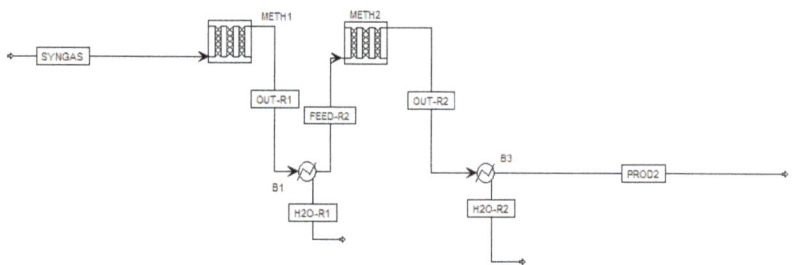

Figure 4. Aspen simulation model showing a two-stage methanation process with intermediate product gas cooling.

- Technical evaluation parameters

To assess the overall performance of the different investigated methanation strategies, several technical evaluation parameters were employed. The CO_x conversion given as a percentage can be specified for a single stage or for the overall process, and is characterized by Equation (12). Generally, the molarities, $n_{i,j}$, are calculated by the molar flow, $\dot{n}_i$, and wet gas composition, x_{ij}, which themselves are calculated by y_{ij}, the dry gas composition, with i specifying each component and j specifying feed or product (see Equation (11)):

$$n_{ij} = \dot{n}_i \times x_{ij} \text{ with } x_{ij} = y_{ij} \times (1 - x_{H_2O}) \tag{11}$$

The CO_x conversion is determined as follows (Equation (12)):

$$U(CO_x) = \frac{(n_{CO} + n_{CO2})_{feed} - (n_{CO} + n_{CO2})_{product}}{(n_{CO} + n_{CO2})_{feed}} \times 100\% \tag{12}$$

The hydrogen conversion is important, especially for the assessment of the sub-stoichiometric and over-stoichiometric strategy, and is given according to Equation (13):

$$U(H_2) = \frac{\left(n_{H_2\,feed} - n_{H2\,product}\right)}{n_{H2\,feed}} \times 100\% \tag{13}$$

For large-scale methanation applications, gross assessment was conducted for all syngas compositions depicted in Table 3, considering only technical feasible methanation strategies. The required hydrogen and the resulting electrolysis capacity, as well as the generated synthetic natural gas output, were ideally calculated and rated to the biomass input power. The evaluation of the overall efficiency of a power-to-gas set-up (Equation (14)) will be discussed in detail in the Results Section.

For large-scale power-to-gas concepts, the required electrolysis capacity was calculated via the required hydrogen amount, complying with the methanation strategies specified in Section 2.1. A specific energy demand for electrolysis of 5 kWh/m³ H_2 was assumed. The capacities of electrolysis and the output of produced SNG were scaled to the biomass input power of 25 MW, while the capacities are always standardized to the lower heating value (LHV). The overall efficiency of the whole process chain from biomass to methane for the considered strategies is defined in Equation (14), with respect to the assumed constant biomass input power of 25 MW, which suitably describes a large-scale scenario [45]:

$$\eta_{overall} = \frac{SNG\ output}{Biomass\ input + Electrolysis\ capacity} \times 100\% \tag{14}$$

The Wobbe Index in kWh/m$^3_{STP}$ (Equation (15)) is defined as the ratio between higher heating value of a gas mixture, H_s, and the root of its relative density, d. More details of each characteristic value can be found in [37].

$$\text{Wobbe Index } Ws \qquad\qquad Ws = \frac{H_s}{\sqrt{d}} \qquad (15)$$

2.3. Assessment Methodology for Each Methanation Strategy

The different methanation strategies from Section 2.1 will be evaluated with the technical evaluation tools from Section 2.2, as described in the following sections. Evaluation parameters from Equations (12)–(15) will be used for the validation, supporting technical assessment tools from Section 2.2 to rate different methanation strategies.

Table 4 provides an overview of which evaluation methods consisting of experiments, modeling and large-scale concept calculations were applied on each specified methanation strategy for the three different syngas compositions. All syngas compositions from Table 1 (steam gasification, SER and 100% CO_2 gasification) were examined based on their potential for downstream direct methanation without additional hydrogen supply in an illustrative ternary plot (C-H-O). Experimental double-stage, lab-scale methanation test runs were only carried out for sub-stoichiometric and over-stoichiometric methanation of syngas from steam gasification. The investigated SN for sub-stoichiometric methanation was identified with Aspen Plus to be 0.78, supplying enough hydrogen so that a full conversion of hydrogen is guaranteed. At over-stoichiometric methanation, SN lies at 1.03.

Table 4. Assessment methodology matrix showing technical figure evaluation tools applied on different methanation strategies and syngas compositions (steam gasification = SG, SER, 100% CO_2 gasification = CO_2-g or none (-)).

Assessment Method	Methanation Strategy		
	Direct	Sub-Stoichiometric	Over-Stoichiometric
Ternary plot	SG, SER, CO_2-g	SG	SG
Lab-scale experiments	-	SG	SG
Modeling approach in Aspen Plus	SG	SG	SG
Evaluation of large-scale PtG concept	SG, SER, CO_2-g	SG, CO_2-g	SG, CO_2-g

The obtained methanation results from laboratory experiments will also be shown in a ternary plot and later compared with modeled results generated in a double-stage methanation process, with implemented reaction kinetics from Rönsch in Aspen Plus. Additionally, all strategies were analyzed based on their suitability for large-scale applications.

3. Results

In this section, the main results from basic thermodynamic investigations for all syngas compositions are presented. Laboratory experimental and simulation results will be shown for two of three methanation operation strategies (sub-stoichiometric and over-stoichiometric), and ideally calculated large-scale power-to-gas set-ups will be assessed based on their technical feasibility for future industrial-scale applications.

3.1. Basic Syngas Composition Evaluation

In Figure 5, Gibbs free energy of the reactions Equations (1)–(9), excluding Equation (7), are shown as a function of temperature. The grey shaded area indicates typical methanation operating temperatures from 250 to 550 °C. Below 250 °C and in the presence of carbon monoxide, poisonous $Ni(CO)_4$ is formed, which blocks active centers on the catalyst surface. Operating temperatures above 550 °C may lead to thermal sintering of the catalyst according to the manufacturer's specifications, which again results in a loss of catalyst activity [46].

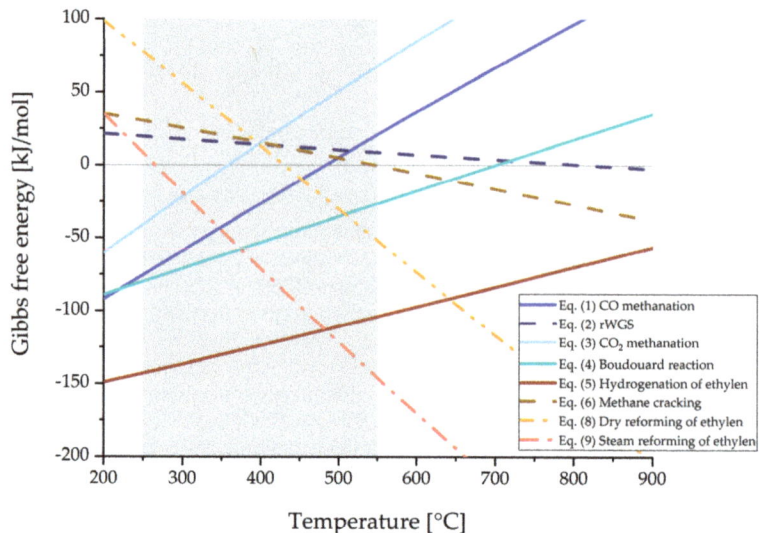

Figure 5. Gibbs free energy for relevant reactions for syngas methanation applications, generated with HSC 10.

Exothermic reactions such as CO and CO_2 methanation (Equations (1) and (3)) happen preferably at lower temperatures, while CO methanation is generally favored against CO_2 methanation as the free energy values are lower. Additionally, ethylene is more likely to be hydrogenated to methane at lower operating temperatures (Equation (5)), which reduces the risk of carbon formation that emerges from methane cracking (Equation (6)), as this process would happen at elevated temperatures. At temperatures from 450 °C upwards, the dominating reactions are the hydrogenation of ethylene (Equation (5)), steam reforming of ethylene (Equation (9)), that again produces a CO/H_2 mixture, and the Boudouard reaction, that raises the possibility of carbon formation. Therefore, temperatures above 550 °C should be avoided in the methanation reactor as the conversion of CO and CO_2 (Equations (1) and (3)) is totally inhibited by all other reactions shown in Figure 5.

3.2. Assessed Methanation Strategies

- Direct methanation (low-investment strategy)

The ternary diagram in Figure 6 depicts different syngas compositions by their C-O-H ratio. The diagram shows that the product composition from SER gasification is not likely to form carbon deposition due to its high share of hydrogen (70 vol.%) in the syngas. The other two syngas compositions (from steam gasification and 100% CO_2 gasification) are located above the carbon deposition line, which would lead to the formation of solid carbon in the case of direct methanation. The reason is a too-low SN, as the values lie at 0.31 and 0.05 for steam and 100% CO_2 gasification, respectively. This result was also proven by Kopyscinski et al. [25], who state that gas produced from coal or biomass gasification offer a too-low hydrogen share for sufficient CO_x conversion and long catalyst lifetime.

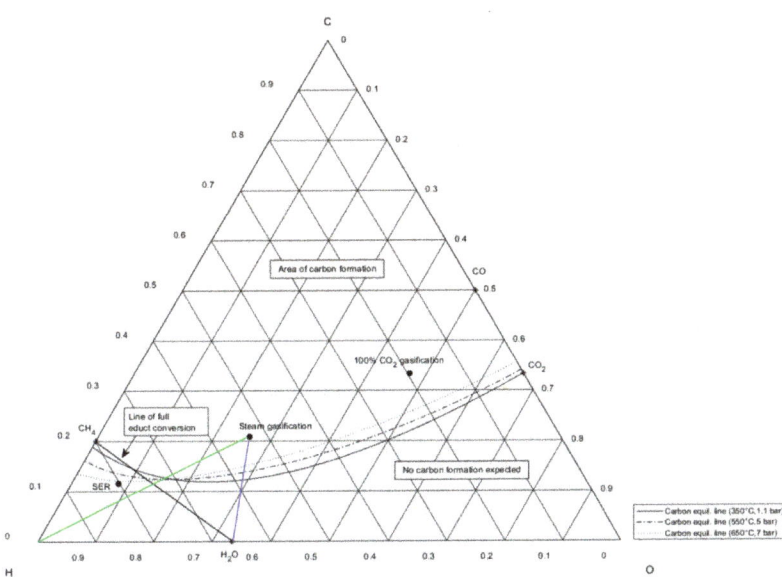

Figure 6. Ternary C-H-O ratio diagram showing carbon equilibrium lines at different temperature and pressure levels and fractional compositions of syngas from Table 1.

A remedial action to suppress carbon formation can be the continuous dosage of water vapor, referring to Neubert [33]. Steam dosage is illustratively shown for steam gasification gas through the blue line in Figure 6 in case of no available hydrogen supply. The optimum operation point for methanation of syngas from steam gasification, which does not lie in the area of possible carbon formation, matches the intersection of the hydrogen supply line in green and the line of full educt conversion in black. Anyway, steam supply would not lead to the desired output gas compositions with a high share of methane as it pulls the C-H-O composition further away from methane on the CH_4-H_2O connection line in the direction of H_2O. Only carbon deposition is hindered through the shift of chemical equilibrium away from the carbon formation equilibrium lines.

The strategy of direct syngas methanation is technologically implementable for small- and large-scale power-to-gas applications only under the usage of SER gasification gas, as its SN of 1.3 shows an over-stoichiometric available hydrogen share of 30%. For the two other syngas compositions, a direct methanation strategy will not yield the desired product gas composition, i.e., a high share of methane, and would immensely shorten catalyst lifetime due to carbon deposition.

- Sub-stoichiometric methanation (maximum H_2 usage)

With two Gibbs reactors applied in the Aspen Plus simulation, the sub-stoichiometric ratio for maximum hydrogen usage was determined as 0.78, as mentioned in Section 2.3. As the least hydrogen should be present in the product gas after a double-stage methanation, a share of unconverted CO_x is expected in the unconditioned SNG. Laboratory experiments with defined dry steam gasification syngas composition from Table 3 were conducted. The results from lab-scale methanation experiments can be seen in Table 5, in which the product gas composition of sub- and over-stoichiometric steam gasification syngas methanation are shown.

Table 5. Results of double-stage methanation experiments of steam gasification syngas in a laboratory plant at the Chair of Process Technology and Environmental Protection in Leoben.

Methanation Strategy (Varying SN)		Sub-Stoichiometric (0.78)	Over-Stoichiometric (1.03)	ÖVGW G B210 Criteria [37]
Final product gas composition				
CO_2	mol.%	16.8	0.01	<2
H_2	mol.%	3.45	13.77	<10
Total CO_x conversion rate after second stage	%	82.5	100	
Total H_2 conversion rate after second stage	%	98.4	94.8	
Combustion characteristics				
Wobbe Index	kWh/m^3	10.86	14.14	13.25–15.81
Higher heating value	kWh/m^3	9.07	9.92	9.87–13.23
Relative density	-	0.69	0.492	0.5–0.7

As hydrogen is restrictively available in the sub-stoichiometric methanation strategy, a high share of 16.8 mol% of CO_2 is present in the product gas of the methanation, which would require CO_2 separation before the product gas can be fed into the natural gas grid in Austria. Although only 3.45 mol% of hydrogen is present in the product gas, meeting Austria's current quality criteria for a feed into the natural gas grid (<10 mol.%), the share of included CO_2 with nearly 17 mol.% is higher than the specified value in the directive. CO_2 can be separated according to today's available technologies (amine scrubber [47,48]) to around 93%. Assuming a moderate CO_2 separation rate of 90% from the raw SNG, the included share of hydrogen rises to about 4.6 mol.% and CO_2 is lowered to around 1.7 mol.%, whereby the gas quality criteria can be met. The hydrogen conversion rate lies at 98.4% due to high reaction temperatures in the first methanation stage, at approximately 460 °C. In the second methanation reactor, temperature peaks were lowered due to limited occurring reactions. The average temperature in the second stage lies at 304 °C, resulting in an elevated hydrogen conversion rate after the first reactor stage.

Figure 7 shows a ternary plot of wet and dry product gas compositions downstream of the first and second methanation stages (R1 or R2), taken from experimental test runs with sub- and over-stoichiometric methanation strategies. The composition of steam gasification syngas is marked in the area of carbon formation, while the connection line of steam gasification syngas with hydrogen in the lower left corner of the plot indicates a methanation process. The line of full educt conversion is shown, connecting the compounds CH_4 and H_2O. The dry product gas from the sub-stoichiometric methanation strategy after the first and second reactor stages is positioned within the area of carbon deposition (blue markers, Figure 7). This very likely results in carbon formation during methanation processes, in small- and in large-scale applications after drying the product gas for a feed into the grid. To hinder unwanted catalyst deactivation, additional steam supply during sub-stoichiometric methanation will shift the equilibrium composition even further towards H_2O, as highlighted in pink markers (Figure 7) for the wet gas composition. This will enable operation in the area below the carbon formation line.

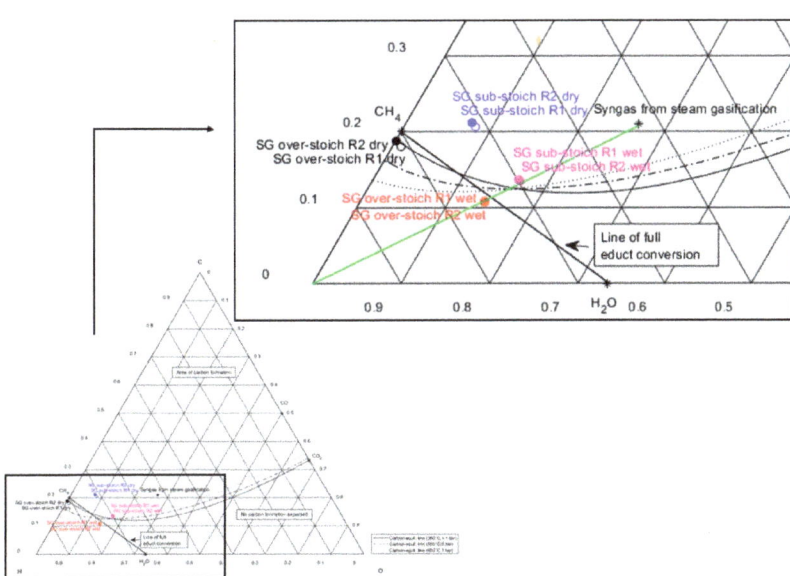

Figure 7. Ternary C-H-O ratio diagram showing wet and dry intermediate and product gas compositions (after first and second stages, R1 and R2) of syngas methanation from steam gasification under sub-stoichiometric and over-stoichiometric hydrogen supply.

- Over-stoichiometric methanation (maximum carbon transformation)

As hydrogen is supplied in an over-stoichiometric ratio during test series, a full CO_x conversion can be achieved for steam gasification syngas methanation. This results firstly in no carbon dioxide present in the product gas (see Table 5). Secondly, the share of left-over hydrogen in the product gas from methanation amounts to nearly 14 mol.%, which demands further gas upgrading to meet the feed-in quality criteria. This can also be assessed by the parameter of lower hydrogen conversion (94.8%).

The experimental results of the over-stoichiometric methanation strategy are also visualized in the ternary plot in Figure 7. The dry gas composition downstream of the first and second methanation stages (black markers) can be found very close to CH_4, because a high share (in this case: 85.1 mol.%) of methane is present in the methanation product gas. For a complete substitution of natural gas, methane concentrations of more than 95 vol.% are required. Otherwise, the product gas can only substitute natural gas to some extent, or needs to be mixed with other gases [33]. Therefore, further gas upgrading after methanation is also necessary for this methanation strategy.

3.3. Summary of Experimental Results of Sub- and Over-Stoichiometric Methanation Strategy

In Table 5, some combustion criteria according to ÖVGW G B210 are listed. In case of sub-stoichiometric methanation, a too-high share of carbon dioxide decreases the higher heating value, that finally does not meet the required criteria. In case of over-stoichiometric methanation, a too-high share of hydrogen is included, whereas the combustion criteria can be met. Further gas upgrading—either H_2 or CO_2 separation—is necessary in both cases for injection of the produced gas into the natural gas grid. Additional costs and energy expenditures for the product gas purification will not be discussed in detail in this article.

3.4. Comparison of Laboratory Experiments and Modeled Results Applying Sub- and Over-Stoichiometric Methanation Strategies

It can be seen in Table 6 that the implemented kinetic model from Rönsch et al. [43] predicts the CO_x conversion behind both methanation stages with a similar trend as the achieved experimental results. The total H_2 conversion in the double-stage methanation process is slightly over-estimated. For sub- and over-stoichiometric methanation, the modeled H_2 conversion lies at 99.5% and 95.5%, respectively. In contrast, it lies at 98.4% and 94.8% according to the experimental results for sub- and over-stoichiometric methanation. Both calculations, CO_x and H_2 conversion, are based on the modeled gas compositions after the first and second reactor stages in Aspen Plus. Additionally, the intermediate (first stage) and product gas compositions (second stage) do not show a major deviation with the obtained gas composition during experimental test runs. Therefore, the kinetic model of Rönsch can describe the experimental results in a good approximation when the system parameters are considered.

Table 6. Experimental and modeled results from sub- and over-stoichiometric methanation strategies.

Syngas from Steam Gasification	SN = 0.78		SN = 1.03	
	1st Stage	2nd Stage	1st Stage	2nd Stage
Dry gas composition (mol.%)—results from experiments/simulation				
CO_2	17.5/16.9	16.81/16	2/3	0.1/0.7
H_2	10.7/12.6	3.45/1	22.6/20.6	13.8/12.1
CO	0.25/0	0/0	0/0	0/0
CH_4	71.55/70.5	79.75/83	75.4/76.4	86.1/87.2
CO_x conversion (%)				
Experimental	80	83	97	100
Modeled	81.5	88.9	99.2	99.6
Total H_2 conversion (%)				
Experimental	-	98.4	-	94.8
Modeled	-	99.5	-	95.5

3.5. Preview: Large-Scale Power-to-Gas Concepts

All syngas compositions shown in Table 1 have been evaluated based on their applicability for large-scale power-to-gas concepts. Assuming 25 MW of constant biomass input power and cold gas efficiencies provided in Table 1, generated syngas power to methanation considering the lower heating value (LHV), the required hydrogen demand and the output of synthetic natural gas have been ideally calculated. To provide an overview about different concept set-ups, the installed electrolysis capacity follows the methanation strategies from Section 2.1 for steam and 100% CO_2 gasification gas. Hydrogen in SER product gas is already present in a widely over-stoichiometric ratio (SN = 1.3), and therefore only direct methanation is assessed for SER product gas. The hypothetical power-to-gas layouts are shown in Table 7.

Table 7. Large-scale power-to-gas concepts following the methanation strategies for the chosen gasification product gas compositions (assumed constant biomass input of 25 MW).

Column Number	1	2	3	4	5	6	7
Gas Type	Steam Gasification			SER	100% CO_2 Gasification		
SN	0.31	0.78	1.03	1.3	0.05	0.78	1.03
Scenario assessed for large-scale applications	(x)	✓	✓	✓	(x)	(x)	(x)
Syngas power to methanation (MW_{LHV})	21	21		18		18	
Syngas volume flow (m^3_{STP}/h)	6909	6909		4670		7762	
Cold gas efficiency gasification (%)	84	84		73		73	
Electrolysis capacity (MW)	0	25	38	0	0	80	107
SNG output [MW_{LHV}] incl.CH_4 in feed	17	30	36	14	8	53	66
Overall efficiency (%)	67.3	59.9	56.9	55.9	30.8	51.1	49.9

Set-ups considering direct methanation of product gas from steam and 100% CO_2 gasification (Table 7, columns 1 and 5) are not technically feasible, as the ternary plot in Figure 6 shows that in these two cases, carbon is certainly formed without additional hydrogen supply during methanation processes. However, in the case of direct steam gasification (column 1), the overall process efficiency would show the highest value of all considered scenarios, with 67.3%, as the included hydrogen share would enable the conversion of CO_x to an appropriate share of synthetic methane (namely, 17 MW_{LHV} SNG). Examining the methanation of CO_2 gasification product gas, a major advantage would be the opportunity for CCU, whereas a very low hydrogen share in the syngas (5 vol.%) leads to enormous electrolysis capacities (80 or 107 MW) for the two chosen methanation strategies (columns 6 and 7). As these capacities are significantly higher than in other investigated scenarios, the overall efficiencies show the lowest values for the methanation of syngas from CO_2 gasification. Recapitulatory, the grey shaded columns in Table 7 show not technically feasible (columns 1 and 5) and not economically feasible (columns 6 and 7) large-scale power-to-gas set-ups.

The columns 2–4 (white background, Table 7) indicate the large-scale power-to-gas concepts which show technical potential for realization.

An immense difference in the required installed electrical power of the electrolysis unit can be seen between the sub- and over-stoichiometric methanation strategies of syngas from steam gasification (columns 2 and 3). In the latter case, electrolysis capacity corresponds to 1.5 times the value of the sub-stoichiometric methanation strategy. In contrast to the high difference in installed power of the electrolyzer, the generated synthetic natural gas power amounts to either 30 or 36 MW for the two methanation strategies. The SNG output power enlargement of 20% underlies the 52% of additionally needed electricity for electrolysis operation, which results in elevated operational costs. The overall efficiency with nearly 60% for the sub-stoichiometric methanation strategy considering maximum hydrogen usage is higher than for the over-stoichiometric methanation strategy with nearly 57% considering maximum carbon usage. At over-stoichiometric methanation conditions, a surplus of hydrogen is produced in the electrolysis, as unused hydrogen (nearly 14 mol.%) can be detected in the product gas, see the experimental results in Table 5.

Due to its high hydrogen share (70 vol.%), SER syngas does not require an electrolysis for stoichiometric methanation of the syngas (column 4, Table 7). However, owing to its low share of carbon dioxide and carbon monoxide (5.6 and 8.6 vol.%), synthetic natural gas with a power of only 14 MW could be produced. Consequently, nearly 38 mol.% of hydrogen is still present in the raw-SNG, which has to be lowered to <10 mol.% to meet Austria's feed-in quality criteria by an appropriate separation step [42] (i.e., polymer membranes). Bartik et al. also indicate around 22 vol.% of hydrogen in the product gas after SER methanation experiments [31]. The overall efficiency of SER product gas methanation lies at nearly 56%, while left-over hydrogen can also be further utilized.

The optimum large-scale set-up is the sub-stoichiometric methanation strategy, which implies the least required hydrogen supply while exhibiting the highest possible overall process efficiency.

3.6. Summary of Results

In Table 8, an overview is provided about the advantages and disadvantages of each investigated methanation strategy applied on different syngas compositions. Additionally, comments are included about why one of the methanation strategies has not been applied on a specific syngas composition. The technological readiness level of the gasification technologies is ranked from top to bottom in Table 8 (steam gasification = best developed technology, SER = medium development status, CO_2-gasification = technology under development).

Table 8. Overview of advantages and disadvantages of the applied methanation strategies on different syngas compositions.

Advantages Disadvantages Comments	Methanation Strategy		
	Direct	Sub-Stoichiometric	Over-Stoichiometric
Steam gasification	- not applicable (SN = 0.31) - solid carbon formation expected	+ maximum hydrogen usage + highest overall PtG efficiency - CO_2 separation from product gas necessary	+ maximum carbon transformation - elevated electrolysis capacity needed - H_2 separation from product gas necessary
SER	+ feasible process + no solid carbon formation + no additional H_2 needed - elevated tar content in syngas	*Not applicable as hydrogen surplus present in raw syngas*	*Not necessary as hydrogen surplus present in raw syngas*
CO_2-gasification	- not applicable (SN = 0.05) - solid carbon formation expected	+ CCU possibility - Low H_2 share in syngas requires large electrolysis capacities	+ CCU possibility - Low H_2 share in syngas requires large electrolysis capacities

4. Discussion

The obtained syngas from gasification features a different gas composition based on the gasification technology (steam or CO_2 as a gasification agent, or SER). Regardless of the gasification technology, all syngases need basic gas cleaning from dust, tar, sulfur, and nitrogen components prior to e-fuel synthesis. The presented results from the combined approach of gasification and methanation proved to be promising, while the different methanation operation strategies considering varying hydrogen supply have a strong effect on the produced SNG gas composition.

Direct methanation represents the best strategy from an economic point of view as the plant complexity and additional expenditures for hydrogen production are minimized. The exclusively suitable syngas for direct methanation is SER syngas due to its high share of hydrogen (around 70 vol.%), as SN lies at 1.31. Included CO_x can be totally converted with the over-stoichiometrically present hydrogen, and carbon formation does not occur. The overall efficiency from the process chain starting from biomass to SNG including SER gasification is the lowest of all considered technically feasible cases in column 4 in Table 7, with almost 56%. According to Fuchs et al. [20], the practical feasibility of the SER process is provided in pilot plant scale. Further investigation should be carried out in a demonstration plant ensuring the working principle in the range of MW and in long-time operation mode. If syngas from steam or CO_2 gasification is directly used for methanation, carbon formation certainly occurs, which would reduce the catalyst's lifetime (see ternary plot in Figure 6). Additionally, conversion rates are not sufficient as the present hydrogen share is low. Both arguments strengthen the reasoning for no further pursuits of a direct methanation strategy if syngas from steam or CO_2 gasification is utilized.

The most promising operation mode is sub-stoichiometric methanation, especially of syngas from steam gasification. Sub-stoichiometric methanation features an optimized hydrogen usage as the availability of renewable energy is restricted for hydrogen production. Laboratory methanation test runs in a double-stage fixed-bed methanation set-up operated with a commercial bulk catalyst showed a 98.4% hydrogen conversion. The pursued aim of maximized hydrogen usage was reached and the overall process efficiency from biomass to SNG was the highest, at 59.9%. Due to too-little hydrogen present, CO_x conversion declined, and the product gas composition did not meet the feed-in quality criteria for the gas grid. Additionally, carbon formation is likely to occur, as it was shown in the ternary diagram (Figure 6). Steam supply may shift the equilibrium composition in the area of no carbon formation. For large-scale sub-stoichiometric biomass to SNG applications, this challenge needs to be precisely considered with fundamental thermodynamic evaluations in the future. The hydrogen supply may be increased, but still appear at sub-stoichiometric character (SN < 1) to firstly save renewable energy and investment costs, and secondly enhance the CO_x conversion. This sweet spot of hydrogen supply can be assessed in future investigations to be a possible optimum operation mode. Replacing the commercial bulk catalyst with a structured honeycomb catalyst that offers better load-flexible operation qualities is also a possibility for further investigations at the laboratory scale [23].

The applied over-stoichiometric methanation showed different results compared with the sub-stoichiometric operation mode. The available excess of H_2 enabled full conversion of CO_x and led to high methane concentrations in the produced SNG, but showed a 94.8% conversion of hydrogen only. Most literature sources focus on this methanation strategy either in fixed or fluidized bed methanation set-ups. A negative aspect of the over-stoichiometric methanation strategy is characterized by the unused share of hydrogen which is present in the produced SNG (around 14 mol.% in experimental investigations by using syngas from steam gasification). Considering an economic point of view, the wastage of elaborately produced hydrogen does not represent a possible operation mode for catalytic methanation of syngas in large-scale applications.

5. Conclusions

The utilization of biomass and excess electricity shows great potential to produce synthetic fuels such as SNG. The focus of this article was on the identification of technological possible methanation routes for a large-scale biomass to SNG set-up, combining biomass-based (b-fuels) and electro-fuels (e-fuels) based on fundamental technical evaluations.

The presented concept, consisting of methanation of syngas from steam gasification, was identified to show the most favorable syngas composition for downstream methanation. The included hydrogen share of around 48 vol.% in the syngas and the high technical readiness level of steam gasification in large-scale gasification set-ups provide a good starting position for further plant concept elaborations including biomass to SNG technologies.

The operation mode of sub-stoichiometric methanation (SN = 0.78) applied to syngas from steam gasification offers a good compromise between the reduction of hydrogen demand and a high overall process efficiency. As the aim is for hydrogen to be used to a maximum extent, laboratory tests showed a hydrogen conversion rate of 98.4%, which represents a satisfactory result. In a large-scale set-up, an overall efficiency from biomass to SNG of 59.9% can be reached. Further studies on predicted catalyst lifetime need to be conducted under the help of water vapor dosage in the methanation system to shift the equilibrium composition into the area of no carbon formation during methanation. The sweet spot of operation needs to be elaborated in detail as the stochiometric number of hydrogen supply and additional steam supply can be varied in future test series. Subsequent gas cleaning after the methanation unit will be inevitable to meet the feed-in quality criteria for H_2 and CO_2.

Author Contributions: Conceptualization, K.S. and T.S.; methodology, K.S. and T.S.; validation, K.S. and T.S.; formal analysis, K.S. and T.S.; investigation, K.S.; resources, K.S. and T.S.; data curation, K.S.; writing—original draft preparation, K.S.; writing—review and editing, K.S. and T.S.; visualization, K.S.; supervision, T.S.; project administration, K.S and T.S. Both authors have read and agreed to the published version of the manuscript.

Funding: This research received no external funding.

Institutional Review Board Statement: Not applicable.

Informed Consent Statement: Not applicable.

Data Availability Statement: Not applicable.

Conflicts of Interest: The authors declare no conflict of interest.

List of Symbols

The following symbols are used in the manuscript:
Index i gas component in mixture
Index j feed or product gas stream
d density
ΔG Gibbs free energy
ΔH^R reaction enthalpy
H_s higher heating value
n_{ij} molarities
$\dot{n}_i$ molar flows
U conversion rate
x_{ij} wet gas composition
y_{ij} dry gas composition
Ws Wobbe Index
η overall process efficiency

Abbreviations

The following abbreviations are used in this manuscript:
AUT   Austria
CCU   Carbon capture and utilization
CH_4 Methane
CO_2 Carbon dioxide
CO_2-g. CO_2 gasification
DFB   Dual-fluidized bed gasification
GHSV   Gas Hourly Space Velocity (h^{-1})
H_2 Hydrogen
H_2O Water or Steam
LHV   Lower heating value (MW)
mol.%   molar share
Multi-T   Multi-Thermocouple
ÖVGW   Österreichische Vereinigung für das Gas- und Wasserfach—Austrian Association for Gas and Water
R1-R2   Reactor 1 or 2
SER   Sorption-enhanced reforming
SG   Steam gasification
SNG   Synthetic natural gas
SWE   Sweden
TU Wien   Technical University of Vienna
vol.%$_{db}$ share in volume percent (dry basis)
wt.%   share in weight percent

References

1. Biollaz, S.M.; Schildhauer, T.J. (Eds.) *Synthetic Natural Gas from Coal, Dry Biomass, and Power-to-Gas Applications*; John Wiley & Sons Inc.: Hoboken, NJ, USA, 2016.
2. Kirkels, A.F.; Verbong, G.P.J. Biomass gasification: Still promising? A 30-year global overview. *Renew. Sustain. Energy Rev.* **2011**, *15*, 471–481. [CrossRef]
3. Hofbauer, H. Biomass Gasification for Electricity and Fuels, Large Scale. In *Encyclopedia of Sustainability Science and Technology*; Springer: New York, NY, USA, 2012; pp. 1426–1445.
4. Kaltschmitt, M.; Hartmann, H.; Hofbauer, H. *Energie aus Biomasse: Grundlagen, Techniken und Verfahren*, 2nd ed.; Springer: Berlin, Germany, 2009.
5. Knoef, H.; Ahrenfeldt, J. *Handbook Biomass Gasification*; BTG Biomass Technology Group B.V.: Enschede, The Netherlands, 2012; Available online: http://www.btgworld.com/en/references/publications/paper-handbook-biomass-gasification.pdf (accessed on 25 May 2021).
6. Müller, S.; Stidl, M.; Pröll, T.; Rauch, R.; Hofbauer, H. Hydrogen from biomass: Large-scale hydrogen production based on a dual fluidized bed steam gasification system. *Biomass Conv. Bioref.* **2011**, *1*, 55–61. [CrossRef]
7. Gil, J.; Caballero, M.A.; Martín, J.A.; Aznar, M.-P.; Corella, J. Biomass Gasification with Air in a Fluidized Bed: Effect of the In-Bed Use of Dolomite under Different Operation Conditions. *Ind. Eng. Chem. Res.* **1999**, *38*, 4226–4235. [CrossRef]

8. Zhang, Y.; Li, B.; Li, H.; Liu, H. Thermodynamic evaluation of biomass gasification with air in autothermal gasifiers. *Thermochim. Acta* **2011**, *519*, 65–71. [CrossRef]
9. Cheng, Y.; Thow, Z.; Wang, C.-H. Biomass gasification with CO_2 in a fluidized bed. *Powder Technol.* **2016**, *296*, 87–101. [CrossRef]
10. Stec, M.; Czaplicki, A.; Tomaszewicz, G.; Słowik, K. Effect of CO_2 addition on lignite gasification in a CFB reactor: A pilot-scale study. *Korean J. Chem. Eng.* **2018**, *35*, 129–136. [CrossRef]
11. Jeremiáš, M.; Pohořelý, M.; Svoboda, K.; Manovic, V.; Anthony, E.J.; Skoblia, S.; Beňo, Z.; Syc, M. Gasification of biomass with CO2 and H2O mixtures in a catalytic fluidised bed. *Fuel* **2017**, *210*, 605–610. [CrossRef]
12. Valin, S.; Bedel, L.; Guillaudeau, J.; Thiery, S.; Ravel, S. CO2 as a substitute of steam or inert transport gas in a fluidised bed for biomass gasification. *Fuel* **2016**, *177*, 288–295. [CrossRef]
13. Pio, D.T.; Tarelho, L.A.C. Industrial gasification systems (>3 MWth) for bioenergy in Europe: Current status and future perspectives. *Renew. Sustain. Energy Rev.* **2021**, *145*, 111108. [CrossRef]
14. Fuchs, J. Verfahrenscharakteristika von Sorption Enhanced Reforming in einem einem Fortschrittlichen Gaserzeugungssystem: Verfahrenscharakteristika von Sorption Enhanced Reforming in einem einem Fortschrittlichen Gaserzeugungssystem, Wien. Available online: https://repositum.tuwien.at/handle/20.500.12708/16812?mode=full (accessed on 27 May 2021).
15. Thunman, H.; Seemann, M.; Vilches, T.B.; Maric, J.; Pallares, D.; Ström, H.; Berndes, G.; Knutsson, P.; Larsson, A.; Breitholtz, C.; et al. Advanced biofuel production via gasification-lessons learned from 200 man-years of research activity with Chalmers' research gasifier and the GoBiGas demonstration plant. *Energy Sci. Eng.* **2018**, *6*, 6–34. [CrossRef]
16. Rehling, B.; Hofbauer, H.; Rauch, R.; Aichernig, C. BioSNG—process simulation and comparison with first results from a 1-MW demonstration plant. *Biomass Conv. Bioref.* **2011**, *1*, 111–119. [CrossRef]
17. Wilk, V.; Hofbauer, H. Analysis of optimization potential in commercial biomass gasification plants using process simulation. *Fuel Process. Technol.* **2016**, *141*, 138–147. [CrossRef]
18. WienEnergie. Available online: https://positionen.wienenergie.at/projekte/mobilitaet/gruener-treibstoff/ (accessed on 1 July 2021).
19. Mauerhofer, A.M.; Müller, S.; Bartik, A.; Benedikt, F.; Fuchs, J.; Hammerschmid, M.; Hofbauer, H. Conversion of CO_2 during the DFB biomass gasification process. *Biomass Conv. Bioref.* **2021**, *11*, 15–27. [CrossRef]
20. Fuchs, J.; Schmid, J.C.; Müller, S.; Hofbauer, H. Dual fluidized bed gasification of biomass with selective carbon dioxide removal and limestone as bed material: A review. *Renew. Sustain. Energy Rev.* **2019**, *107*, 212–231. [CrossRef]
21. Asadullah, M. Biomass gasification gas cleaning for downstream applications: A comparative critical review. *Renew. Sustain. Energy Rev.* **2014**, *40*, 118–132. [CrossRef]
22. DNV GL. *Methanation: Technical Fundamentals and Market Overview*; Report Methanation (vs 1 dec 2019), Report number: OAG.19.R.10157194; DNV GL: Bærum, Norway, 2019.
23. Lehner, M.; Biegger, P.; Medved, A.R. Power-to-Gas: Die Rolle der chemischen Speicherung in einem Energiesystem mit hohen Anteilen an erneuerbarer Energie. *Elektrotech. Inftech.* **2017**, *134*, 246–251. [CrossRef]
24. Bundesministerium für Verkehr, Innovation und Technologie und Bundesministerium für Nachhaltigkeit und Tourismus, #mission2030—Die Österreichische Klima- und Energiestartegie. Available online: https://www.bundeskanzleramt.gv.at/dam/jcr:903d5cf5-c3ac-47b6-871c-c83eae34b273/20_18_beilagen_nb.pdf (accessed on 28 May 2021).
25. Kopyscinski, J.; Schildhauer, T.J.; Biollaz, S.M.A. Production of synthetic natural gas (SNG) from coal and dry biomass—A technology review from 1950 to 2009. *Fuel* **2010**, *89*, 1763–1783. [CrossRef]
26. Lehner, M.; Tichler, R.; Steinmüller, H.; Koppe, M. *Power-to-GAS: Technology and Business Models*; Springer: Cham, Switzerland; Heidelberg, Germany, 2014.
27. Rönsch, S.; Schneider, J.; Matthischke, S.; Schlüter, M.; Götz, M.; Lefebvre, J.; Prabhakaran, P.; Bajohr, S. Review on methanation—From fundamentals to current projects. *Fuel* **2016**, *166*, 276–296. [CrossRef]
28. Medved, A. The Influence of Nitrogen on Catalytic Methanation. Ph.D. Thesis, Montanuniversität Leoben, Leoben, Austria, 2020.
29. Gao, J.; Wang, Y.; Ping, Y.; Hu, D; Xu, G.; Gu, F.; Su, F. A thermodynamic analysis of methanation reactions of carbon oxides for the production of synthetic natural gas. *RSC Adv.* **2012**, *2*, 2358. [CrossRef]
30. CHEMCAD: Chemstations. Available online: https://www.chemstations.com/CHEMCAD/ (accessed on 31 May 2021).
31. Bartik, A.; Benedikt, F.; Lunzer, A.; Walcher, C.; Müller, S.; Hofbauer, H. Thermodynamic investigation of SNG production based on dual fluidized bed gasification of biogenic residues. *Biomass Conv. Bioref.* **2021**, *11*, 95–110. [CrossRef]
32. Bartholomew, C.H. Mechanisms of catalyst deactivation. *Appl. Catal. A Gen.* **2001**, *212*, 17–60. [CrossRef]
33. Neubert, M. Catalytic Methanation for Small-and Mid-Scale Sng Production. Ph.D. Thesis, Friedrich-Alexander Universität Erlangen-Nürnberg, Erlangen, Germany, 2020. Available online: https://opus4.kobv.de/opus4-fau/frontdoor/index/index/start/2/rows/20/sortfield/score/sortorder/desc/searchtype/simple/query/neubert+michael/doctypefq/doctoralthesis/docId/13118 (accessed on 2 June 2021).
34. Kienberger, T. Methanierung Biogener Synthesegase Mit Hinblick Auf Die Direkte Umsetzung von Höheren Kohlenwasserstoffen. Ph.D. Thesis, Institut für Wärmetechnik, TU Graz, Graz, Austria, 2010. Available online: https://www.osti.gov/etdeweb/biblio/21397329 (accessed on 2 June 2021).
35. Wang, S.; Bi, X.; Wang, S. Thermodynamic analysis of biomass gasification for biomethane production. *Energy* **2015**, *90*, 1207–1218. [CrossRef]

36. Tremel, A.; Gaderer, M.; Spliethoff, H. Small-scale production of synthetic natural gas by allothermal biomass gasification. *Int. J. Energy Res.* **2013**, *37*, 1318–1330. [CrossRef]
37. VGW Österreichische Vereinigung für das Gas- und Wasserfach. *Erdgas in Österreich—Gasbeschaffenheit ÖVGW G B210: 2019 06*; ÖVGW Österreichische Vereinigung für das Gas- und Wasserfach: Wien, Austria, 2021.
38. Thema, M.; Bauer, F.; Sterner, M. Power-to-Gas: Electrolysis and methanation status review. *Renew. Sustain. Energy Rev.* **2019**, *112*, 775–787. [CrossRef]
39. Krammer, A.; Medved, A.; Peham, M.; Wolf-Zöllner, P.; Salbrechter, K.; Lehner, M. Dual Pressure Level Methanation of Co-SOEC Syngas. *Energy Technol.* **2021**, *9*, 2000746. [CrossRef]
40. Frick, V.; Brellochs, J.; Specht, M. Application of ternary diagrams in the design of methanation systems. *Fuel Process. Technol.* **2014**, *118*, 156–160. [CrossRef]
41. Bai, X.; Wang, S.; Sun, T.; Wang, S. Influence of Operating Conditions on Carbon Deposition Over a Ni Catalyst for the Production of Synthetic Natural Gas (SNG) from Coal. *Catal. Lett.* **2014**, *144*, 2157–2166. [CrossRef]
42. Kirchbacher, F.; Biegger, P.; Miltner, M.; Lehner, M.; Harasek, M. A new methanation and membrane based power-to-gas process for the direct integration of raw biogas—Feasability and comparison. *Energy* **2018**, *146*, 34–46. [CrossRef]
43. Rönsch, S.; Köchermann, J.; Schneider, J.; Matthischke, S. Global Reaction Kinetics of CO and CO_2 Methanation for Dynamic Process Modeling. *Chem. Eng. Technol.* **2016**, *39*, 208–218. [CrossRef]
44. Klose, J. Kinetics of the methanation of carbon monoxide on an alumina-supported nickel catalyst. *J. Catal.* **1984**, *85*, 105–116. [CrossRef]
45. Larsson, A.; Kuba, M.; Vilches, T.B.; Seemann, M.; Hofbauer, H.; Thunman, H. Steam gasification of biomass—Typical gas quality and operational strategies derived from industrial-scale plants. *Fuel Process. Technol.* **2021**, *212*, 106609. [CrossRef]
46. Schmidt, M.; Schwarz, S.; Stürmer, B.; Wagener, L.; Zuberbühler, U. *Technologiebericht 4.2a Power-to-Gas (Methanisierung Chemisch-Katalytisch) Innerhalb des Forschungsprojektes TF_Energiewende*; ZSW: Stuttgart, Germany, 2018.
47. Aaron, D.; Tsouris, C. Separation of CO_2 from Flue Gas: A Review. *Sep. Sci. Technol.* **2005**, *40*, 321–348. [CrossRef]
48. Fischedick, M.; Görner, K.; Thomeczek, M. (Eds.) *CO_2: Abtrennung, Speicherung, Nutzung*; Springer: Berlin/Heidelberg, Germany, 2015.

MDPI
St. Alban-Anlage 66
4052 Basel
Switzerland
Tel. +41 61 683 77 34
Fax +41 61 302 89 18
www.mdpi.com

Energies Editorial Office
E-mail: energies@mdpi.com
www.mdpi.com/journal/energies

www.ingramcontent.com/pod-product-compliance
Lightning Source LLC
LaVergne TN
LVHW070545100526
838202LV00012B/381